User Group Leadership

Michelle Malcher

Apress®

User Group Leadership

ISBN-13 (pbk): 978-1-4842-1116-8

ISBN-13 (electronic): 978-1-4842-1115-1

Managing Director: Welmoed Spahr
Lead Editor: Jonathan Gennick
Editorial Board: Steve Anglin, Pramila Balen, Louise Corrigan, Jim DeWolf, Jonathan Gennick, Robert Hutchinson, Celestin Suresh John, Michelle Lowman, James Markham, Susan McDermott, Matthew Moodie, Jeffrey Pepper, Douglas Pundick, Ben Renow-Clarke, Gwenan Spearing
Coordinating Editor: Jill Balzano
Copy Editor: Laura Lawrie
Compositor: SPi Global
Indexer: SPi Global
Artist: SPi Global

Distributed to the book trade worldwide by Springer Science+Business Media New York, 233 Spring Street, 6th Floor, New York, NY 10013. Phone 1-800-SPRINGER, fax (201) 348-4505, e-mail orders-ny@springer-sbm.com, or visit www.springeronline.com. Apress Media, LLC is a California LLC and the sole member (owner) is Springer Science + Business Media Finance Inc (SSBM Finance Inc). SSBM Finance Inc is a **Delaware** corporation.

For information on translations, please e-mail rights@apress.com, or visit www.apress.com.

Apress and friends of ED books may be purchased in bulk for academic, corporate, or promotional use. eBook versions and licenses are also available for most titles. For more information, reference our Special Bulk Sales–eBook Licensing web page at www.apress.com/bulk-sales.

Any source code or other supplementary material referenced by the author in this text is available to readers at www.apress.com.For detailed information about how to locate your book's source code, go to www.apress.com/source-code/.

*For my extraordinary friends who dependably provide
encouragement and support*

Contents at a Glance

Contents

About the Author

Michelle Malcher (@malcherm) is a well-known volunteer leader in security and database communities. She is currently on the board of directors for FUEL Palo Alto Network User Group, and has served on the board of directors for Independent Oracle User Group (IOUG). As an Oracle ACE director, she has had the opportunity to present around the world at user group conferences. She continues to participate on the conference committee for the IOUG. Her technical expertise on topics from database to cybersecurity, as well as her senior-level contributions as a speaker and author, have aided many corporations across architecture and risk assessment, purchasing and installation, and ongoing systems oversight. These corporations include Wells Fargo & Company, where Michelle is currently team lead for Enterprise Database Security.

Acknowledgments

Believe it or not, there is a tweet out there stating that I would not be writing another book. Of course, my user group friends did not believe me, and with this book, they were right. Writing articles and books is not an easy task, and technical books I have always wanted to make sure to go above and beyond to add value and relevance. It does take quite a bit of time. However, I have a wonderful support in user group communities. It is my involvement with these fantastic user groups that made writing this book so much easier. I wanted to share my experiences so that others could hopefully be a part of a community.

I am very much appreciative of all of the friendships I have worldwide that are part of the Oracle user community and cybersecurity user group. In sharing my experiences, I not only have been able to do my job better but also provide way to continue to share and mentor others. It is very difficult to acknowledge only a few people because of group of people that have encourage and support over the years.

I would like to thank the previous and current board members and other support staff of the Independent Oracle User Group and FUEL. They believed in me enough to mentor me and to allow me to grow in leadership skills. They demonstrated ways to work together to develop the community. Thanks also to all of my friends in the Oracle ACE program and the support staff. It is difficult to thank just a few people because in networking in the community you are touched by several people in different areas which is simply wonderful. Thanks!

I normally thank my junior DBAs, Amanda and Emily, but in this case I also will acknowledge them for their leadership skills that they demonstrate in school and other activities. I am proud of them and enjoyed having them sit next to me while writing as Amanda did homework and Emily wrote some short stories.

Thank you to all of you reading this book. I hope that it inspires you and fuels your passion for the user group community.

Foreword

The function of a user group has never been as vital as it is today, to all parties it unbiasedly serves. This includes the individual technologist, the technology provider, and the organizations consuming and benefitting from the technology.

What users want from a user group coexists well with what a technology provider wants from the user community, especially if it's a healthy and thriving user-led community. User group members want access to education, product knowledge, training, a network of experts and professionals, and answers to their questions. Technology providers want the ability to continually learn about customer needs, keep tabs on markets, influence buyers, accelerate buying decisions, and receive quality and timely feedback. All of this is only possible within a user group.

In today's "yelp.com" world, where technology comparisons and opinion sharing are highly valued, user groups are more important than ever as they provide a difficult to replicate platform of timely, unique, trusted, and unfettered content and context. In this way, a user group functions as a nonthreatening, critical hub that connects buyers and sellers, users and developers, and customers and vendors.

Michelle Malcher has written an excellent book as a resource for anyone wanting to get involved and make a difference. It's a must-read for anyone aspiring to a leadership role in a group. It's good reading for just the typical member, too, because someday you'll aspire to greater things.

—Josh Berman
Director, SmithBucklin Technology Industry Practice
www.SmithBucklin.com

■ ■ ■

CHAPTER 1

Development of a User Group

One of the best career decisions I have ever made was becoming a member in a user group, and probably the best career decision was volunteering for a user group.

As you can tell, I am a strong believer in user groups and the advantages that they provide. The opportunities in the development of a user group are incredible: bringing together a community of like-minded individuals for education, networking, and working with the vendor. I get excited about the knowledge sharing and possibilities of problems being solved and new solutions being developed. In this chapter we will discuss the development of a group, with additional information to come in future chapters about building leaders and volunteers. These discussions should provide you with what is needed to create and build a group.

User groups are important for the users in a community for networking, growth, and learning. They are important for the vendors because there is an active community providing feedback. We cannot forget to mention the group of users that they now have engaged and already interested in their products. There are mutual benefits to the company and the users in the community and we are just scratching the surface here of a few of the benefits.

Building a community of common users and professionals is an excellent way to gain valuable knowledge and develop a career that is rewarding and enjoyable in an area of interest. There are plenty of other soft skills that are picked up along the way, with the benefit of additional learning and opportunities. There is extra effort in being involved and starting the group but the benefits far outweigh the work and time put into the group. Investments benefit both sides and the development of a user group will be able to be successful if both are seeing the value of their investments.

Reasons to Begin

There are four reasons that drive an interest to create a user group. In addition to having enthusiastic users and volunteers, the mission of the group centers around one of these four reasons. If goals or mission is not part of the development, it will be difficult to gather and build the community. A common thread will allow for the group to be focused and developed around the same reason. The four threads that we will discuss in the rest of this chapter are:

- Common Purpose
- Common Problem
- Common Need
- Culture

It is possible that some of these reasons have overlap, but the main focus will fall under one of these with some possible additional goals that pull in the overlapping areas. Starting here is important to make sure that the group will be able to be a viable long-lasting group. The development of a group might be a difficult process and a common thread will keep the movement going.

A Common Purpose

Technology, area of interest, a vendor, a company, and industry are all common purposes for a group to be developed. It might be a group of technology professionals wanting more from the technology or even just excited about the technology that they are working with and wanting to see it grow and advance.

Database management and security are my areas of passion. I tend to spend time outside of working hours gathering more information and researching on my own. There are LinkedIn groups I am part of in these areas, and user groups where I am a leader and member of the community. I am not alone in this area; there are other groups out there with the same area of interest. The goal would be to bring a group together that can have this common purpose, which might be to advance knowledge, develop careers, and provide growth in a new or existing area.

A common purpose does not have to come from just a group of individuals. A company could have the purpose of providing the customer base with an excellent experience, and it is important to recognize that a user community can assist in developing this strategy. From the company perspective, there is a wide set of advantages to polling the customer base and providing information to a user group. A company might be the one even developing the group or it starts to evolve out of the customer base. Even with it coming from the company, there might be an opportunity to pull this out to be an independent group. Discovering that development of the user group is moving outside of vendor control would be meeting the purpose to form an independent group.

There is a sense of passion and drive around a purpose and meaning for the group, for example, wanting this area to grow and succeed or to be the best in the industry. Purpose does focus on a main area, which probably makes it easier to articulate a mission and goal to bring a community together.

A Common Problem

Developing a group can stem around a common problem that needs to be addressed. In looking for an answer to a problem or troubleshooting some gaps or weaknesses in product, discussions between the community and vendor are very important. An isolated problem is not useful for driving people together, but a problem like software installation failures on a common platform or limited documentation which causes one to develop their own process just to be able to complete the installation would be a more general problem.

Another example of a common problem would be lack of a secure configuration, which would be expected by default. If the users of the software take these problems back to the vendor, the vendor has the opportunity and support of a community to take a look at the problem and then to help validate and test that the problem has been resolved. This is also an area where the vendor can step in and provide support and develop an inside community to look at these important issues.

In developing a user group around a common problem, what happens when the problem is solved? Do you hope for another problem to rally around? Or does the group dissolve after the resolution of the problem. If you are to put the effort and energy in creating the group, it would be a disappointment if there was not a reason to continue. A common problem will need to lead into another purpose or mission statement for the group, but more likely it can lead to a common need, especially if you consider the problem around a secure configuration, and how this could develop into general security and review as there are new releases and other opportunities to support these needs.

A Common Need

The group can develop out of a need from the users or vendor. A need can come out of solving a problem or a need might be part of the problem, but a need also can be much more. It might be the ability to adapt to an ever-changing environment in planning for the future, or a need to have enhancements to a product and services.

Sometimes one company can have specific needs or have a unique situation that does not necessarily relate to others, but this is rare. Even if thinking the situation is unique, it might not be; it is usual that this involves more than one group of people or a company. The community will only be able to grow around a need if it does reflect one common to the industry, something that will support the users better or help plan for the future.

A common need might be education provided by the users. In understanding how technology is used by other companies, the information from other users can provide examples of what others are doing. Best practices, standards, designs, and configurations are all excellent areas to educate others and share ideas. Especially in the technology fields, topics such as upgrades, patching, new features in the releases, and how the environments are configured are top education choices that user community can provide. Technology professionals are continuously learning to deal with the complex environments and ever-changing technology, and education is high on the list for joining a user group. Networking is close behind. However, we are getting ahead of ourselves; we will discuss networking in later chapters with strategies and planning.

But before we continue with the other common thread, let's finish up the need from the vendor perspective. The vendors are attempting to sell their products, and even if they provide excellent information, where else can someone validate that but against the users of the products? Working with the users in the community to provide education and training and share ideas satisfies several needs in other companies and users. The developers of the products might be looking for customer features and enhancement and they are trying to figure out how to tap into others to get their experiences. The need to support the customers and receive feedback like this will drive interest and allow for growth in a user community.

Culture

Culture is an important connection to discuss because it will give a base of how a user group can develop or grow. If the culture is one that keeps silent and more focused internally, it will be more difficult to have a community. The culture needs to be one that is more open and willing to discuss and share—of course, not sharing company secrets or intellectual property, but willing to learn and share from others. Additional considerations should be examined to support a user group, such as is there a following already around the technology or product, or is there a grassroots group already forming, and having this basic level of activity for the users supports the development of the user group and has already started to form the culture of the group.

There is an opportunity to leverage existing culture to grow and cultivate a user group. The culture might exist because of the nature of the vendor, it could be from a new technology coming from the schools and the students graduating are bringing the culture with them. It can also be industry culture and how the industry is involved in bringing the behaviors and characteristics from the environment.

The culture can be leveraged to expand a user community, by understanding if one already exists, and the dynamic of that culture can bring users together. A culture does not have to be all like-minded individuals nor do all users have to have the same problem or need; they can have diversified issues and come from different interests with the same common belonging that identifies them as part of the group.

An example of a culture is database administrators (DBA). The DBA has responsibility for the company data and also supports various departments and business units within the company. DBA jobs from company to company can be very similar. There are DBAs handling different roles in the company with backups to working with developers and application design. So there is some diversity in the role and how different DBAs approach their jobs. However, a group of DBAs from different companies brought together can understand and relate to each other because of their commonalities. There is a culture among DBAs because of these relationships. There might even be subcultures or their own cultures with database platforms. The Microsoft SQL Server DBAs might rally around the product as would the Oracle DBAs around their databases. These are great examples of existing cultures because of common job responsibilities and product relationships.

A couple of other cultures focus on the cloud and security. There are cultures that do exist around security professionals again because of the nature of the tasks and functions they perform. These cultures of professionals provide occasions to evolve into user groups either based on the emerging technologies or on the topic itself.

Leveraging the existing cultures and noticing that there is already a following in a particular area increases the growth potential for the user group. However, you need to also understand that if a culture is not available to share or participate in a user group it will also limit the value and potential for that group to develop. Even if the culture does not currently exist, it will be needed as an element of the user group.

Development

Common areas and threads that can support a user group, but this is not all that is needed to develop one. Threads and common areas are essential for the success of the community. As seen in Figure 1-1, the areas we just discussed can overlap, and there can be a combination of these to develop into a user group. At a minimum, one of these areas is required in order to have volunteers and the energy to develop the group. Multiple areas just strengthen the opportunity to grow and demonstrate the importance of having a group in place.

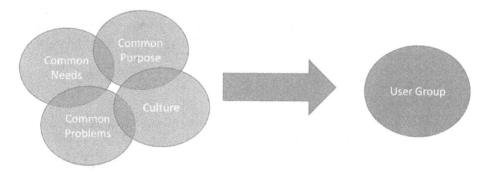

Figure 1-1. *Common Threads*

Now that we have established that there is an opportunity here for a group to have a mission and a goal because of these commonalities, development of the user group should follow leveraging the people who are already involved and interested. It is important not to try this alone, unless you have quit your job and can do this full time or have found a way to have 30 hours in a day. There is a great deal of effort to get a group going and sharing the workload is important, especially as it is on top of regular work.

This is the time to round up the volunteers and get them ready to begin the user group and start up the community. Getting started with the development involves working through the following steps:

1. Verify if there are other groups in this space.

2. Decide on location of this group, online, regional, or international.

3. Develop mission and goals of the user group.

4. Research association help and staff needs.

5. Plan vendor sponsorship or other sources of support.

Verify Other Groups

If there are other user groups in this space, does it make sense to start a new one? Maybe or maybe not, but it does depend on the focus area and if it considered part of an existing group. A new group can be considered for a different location, industry, or focus area. It is definitely worth a little research to understand if there are other choices for the community. If there are existing groups, is there an opportunity for a different location or industry? For example, if you are looking to start a user group of Oracle DBAs, there is a group that already exists, the Independent Oracle User Group (IOUG); however, there might be a need for a regional group or a need for leadership in one of the regions. Developing a user group can be based on another existing group in a different region or location.

If another database platform community might not fit in the IOUG, then a group can be developed based on that platform. It is important to realize that even if one database user group exists, the culture, purpose, needs, and problems might all be different with this community. Recognizing differences will allow the group to flourish and grow in the way that it needs to.

There is a lot of excitement and energy in creating a new community, but if a similar group exists, adding the energy to the existing community might make sense, too. If not, the next planning steps continue with the development of the group.

Group Location

The user group can be a strictly online and virtual group or support an in-person meeting type of community. There have been quite a few groups that have started up as hybrids, using an online community for year-round activity, involving the membership in discussions and information, and then holding in-person networking events. There are varying costs associated with the different options and activities that a community can participate in, which might influence the type of community that is developed.

For starting up, there might be a couple of group meetings to get a few people involved. An easy way to do this is through meetups and gathering at a restaurant for lunch or drinks after work. This will start the networking and planning. The user group will be able to establish interest and focus, and having a few or more people to meet with will help in deciding some direction.

There is a way to get started with informal meetings and still address some of the common threads while gathering volunteers and planning strategy for a larger group. The membership can support these type of meetups, but vendors or product sponsors can help out with other type of events and also online support.

In later chapters, we will discuss in-person meetings and virtual presence, but for now, look at the culture of the group. Is this going to be one with more of a social media presence, with questions posted in discussion rooms? Or is the culture more of an in-person one, with members who would rather sit in front of a white board, carrying on discussions face-to-face with refreshments? Or might it be a combination of both?

These are the main questions for understanding location and what is meant about the user group and how the meetings should be taking place. The types of meetings can still be different, but looking at the purpose of the group and the culture, it helps to look at the location for developing a user group that meets these needs.

Develop a Mission and Goal

Even if the development of this group stems from a common purpose, developing a mission statement and a set of goals is needed for beginning the group. The purpose defines why the group is coming together, but the goals give an opportunity to review the purpose and accomplishments for the year and adjust as needed. A mission provides a clear way to talk about the user group to plan what will be needed and enable volunteers to help out.

The development of the user group might survive on just a common purpose, problem, or need without a mission statement, but to be able to articulate what the group is about and communication of these goals is what allows others to embrace the community and become part of it. It is this communication of the mission of the group that will allow for the focus of the activities.

Sometime it is the easiest part of developing a user group to create a mission statement, which means that there is clear meaning behind the group. Other times, it is not necessarily the clear meaning but the growth of the group, trying to take on different activities, that might want the mission expanded. Technologies change quickly and technology companies can be acquired faster than in some other industries. As these groups build upon each other, this is when the goals of the group might shift or changes in the mission statement will need to be adjusted.

To create a mission statement, keep it simple: state the composition of the group and why it exists. The group being developed might be technology professionals focused on data, and if they are looking to educate and network with others interested in data analytics, then say so.

Research Possible Association Management

Development of a group by yourself is very difficult. There are management companies that can assist in planning meetings and all things for the user group. There are many tasks that need to be accomplished. If this is a very popular area a user group community has the potential of growing very quickly and as a volunteer doing this part-time, it can be very difficult to manage and bring together all of the pieces necessary for the development of the group. Is this a volunteer position or your full-time job? If you are just a volunteer, you can always have three to four times more volunteers more than what you think you need or research association management companies.

Management companies can provide additional staff to support a user group, help manage events and a website, and even help with budget planning. They normally charge fees based on time and staff that are needed. They provide services for a large variety of user groups, and depending on the size of the group can leverage different experiences and venues to help with planning and meetings.

Volunteers can also help get things off the ground, but unless the user group is for association management, volunteers might have some gaps in information. This book prepares you to know how to develop a group and will give some very practical experience about volunteers and enabling leaders, as well as some important planning tips. Some extra research on association management will help with some of the logistics and even look into some of the staff options to implement some of the plans for the user group.

Plan Vendor Sponsorship

Vendors want to get in front of user communities. These are the groups of people deciding on what products to purchase and understand their environments to talk about what is working and is not. Vendors have marketing dollars to sponsor event or conference booths. The mission statement and the reach of the community are important to attract vendor sponsorships.

The user group might revolve around a product and a vendor. The vendor might want to support the startup of the user group, through sponsor funds for the website, through planned events, or through sponsorship of meetings. This vendor could have exclusive rights then to sponsor the group, depending on the agreements and control of the user group.

There are several vendors that are in the environment and complement each other, and having several sponsors add value to the users to be able to get information about the different vendors and how they work together. The vendors can also showcase how they work well in these environments. This type of sponsorship might be needed for a conference, and also can provide funds for running webinars and supporting the virtual offerings of the user group.

Asking for sponsorship will be beneficial for both sides: for the vendors to be in front of the participants and for the users to keep informed about changes and products.

Summary

Developing a user group is an excellent way to fuel your passion for the things that you do and enjoy. A group of users have common threads or areas that are important to keep them working together as community. The commonalities of the group will help to grow and thrive as a community. It is usually those overlap areas that allow for the group to think about roadmaps and future events and activities.

A common purpose, need, problem, or culture is an absolute first step to define the user group being developed. From this, a mission and goal of the group is created. Communication of the purpose of the group can be done by using this mission statement. It will encourage volunteers to get involved and show vendors reasons to become engaged in the events. Developing the group means opportunities to develop the users and community through education, networking, and learning through other users.

The energy and effort put into the group will not be wasted and will be used to launch into new career paths or breathe new life into existing ones. There will be challenges and rolling up of sleeves to get the work and planning done, and the following chapters will help you through that to be able to be involved in a group to be excited about.

CHAPTER 2

■ ■ ■

User Group Governance

Governance of a user group might not be one of the first things that you think of when getting started, or even in the leadership skills that are developed as being part of the user group governance, but it is critical for the group to be able to function as a coherent entity. It is possible talking about governance first might be getting ahead of developing of the leaders in the community; however, seeing the goal of what needs to be established supports the need and highlights the areas for growth.

The structure of the user group is defined by the governance, and the governance is defined by the leaders. It is almost a chicken-and-egg situation, but not quite, as there are several examples that can be followed. Governance is not just about the how the group is led; it also describes the culture and lays the groundwork for the future with succession planning. Governance allows for the group of volunteer leaders to come together with the same understanding and structure to collaborate, set up the business strategies, and keep the group focused. Working on a board or a committee is difficult enough and most user group leaders are volunteers and have other outside interests and jobs pulling them in different directions. The structures that are used for governance should take this in consideration. Even though the volunteers are enthusiastic about the user group and have vested interests in contributing to the group, the governance will organize the volunteers and provide the needed guidance to be effective.

The governance framework will be used to set up if the user group will have a board of directors or council to provide leadership, and how committees can be set up. Even if all of the positions cannot be filled right away, having the framework in place will help in seeking volunteers and provide the valuable structure for the organization to function. In the beginning the leaders starting the group might be appointed to the board of directors where down the road the membership can vote on the board members. Deciding to have a board of directors plans for the future of the user group. Even a framework of a user council and how that can develop into a board would be planning ahead, and allow for a simple start up. Governance framework needs to look at the leadership of the group, planning for the future growth, and development of the culture.

Let's look in more detail at the board of directors, committees, leaders, and succession planning.

Boards

The governance framework includes the structure of the board, information such as how many people there are on the board as well as the terms and positions for board members. The leaders should form either a board or council with specific roles. Having this in place gives the needed structure to plan for the future and involves other leaders in succession planning or building out the board. In technology, there are not always opportunities to develop leadership skills and participation on the board gives an excellent way to demonstrate the leadership skills.There are many times day to day to worry about the details of the job, about getting tasks completed. Managing the databases and putting controls and processes in place for the environment is the focus, but there are important soft skills, leadership skills, for project management, that need to be developed. It is also possible that in the technology area there is not the opportunity to demonstrate or use the leadership skills. The board and other volunteer areas of the user group can pull in these technology leaders and allow them to really round out their careers.

Each user group might have a different focus for the positions of the board. Deciding the focus areas will help for the different roles. There might be a conference associated with the user group or even the vendor, so having a role that focuses on events, conferences, and education makes sense. There are definitely vendor relationships. Finance is a primary role, for someone to work on the budget and review the expenses and revenue of the board. As a volunteer organization, the goal is not to make a profit, but expenses are a part of running the group as well as income for memberships and conferences. Membership is another role that is pretty standard because member communication and activity is necessary. Also, look at the mission statement to make sure that the rest of the roles match up. Do you want there to be growth in membership? Then marketing is needed. If you have hot topic areas, is a board member going to make sure that there is a focus in education without drifting area from the main focus of the board? Taking care of volunteers could be a board role. A board member involved in the different groups will help the groups stay aligned with the direction and strategy the board has set.

The board sets the governance, culture, and strategy for the user group. They lead the volunteers and represent the members of the community. SmithBucklin is an association management and services company, and they offer an excellent Leadership Institute program developed and led by Henry Givray, President and CEO of SmithBucklin. Givray has written numerous articles. In one of these articles, he states some of the things that only a board of directors can do:

- Ensure alignment with purpose of the group

- Preserve the community's core values

- Plan the future

- Keeps the organization on track

- Communicate vision, core values, and strategies with the community

- Select and identify standards for leaders and volunteers

- Create and maintain governance framework

The purpose of the group needs to be defined, which is normally an easy step, because it is a user group. The purpose and reason for existing should come from the user responsibilities. The board needs to define the mission and goal and verify that this aligns with that reason for having the group. The board of directors is really the only group to ensure that the plans even if adjusted and change continue to match up with the purpose of the group.

Typically the commonality and culture of the group comes from the job role and vendor that they are using. This culture might be understood, but there is a deeper heart of the group which is the core values. It is up to the board of directors to communicate the core values and keep the strategies in-line with these values. Core values normally do not change and are the underlying truths for the group. However, change seems to be the one constant, and as the focus of the group might change, it is the responsibility of the board to verify a change in focus is still true to the core values.

Planning for the future is something that only the board can do. Managing timing is based on financial and volunteer resources, so this will come from the board as well. Especially with user groups and volunteers, you will find even though there are great ideas and probably a ton of things that are the group is wanting to accomplish, needing to prioritize and allocate the resources are going to be part of this plan for the future.

The board of directors has to guide the community in the right direction and keep the course or steer to make sure that the initiatives are consistent with the mission and meeting the membership needs. It is possible to get distracted with a hot topic and appear to be following a squirrel. The review and discussion in the board room should decide how to handle the new and shiny information to decide where it fits into the priorities and strategy. Technology user groups might have to deal with this more frequently because of the constantly changing technologies, enhancements and development of new technologies. New topic areas can possibly grow into their own user group, but the focus must stay on the mission and core values of the group. The board must perform the task of mapping the future while confirming and alignment of the goals. A shift in direction might be needed to add new members and incorporate a new technology, but this is only something a board can decide. With the board decisions and direction, communication has to filter through to the stakeholders, councils, and members. Only the board can clarify and provide the context for the visions and goals. Transparency by the board will allow others to be more involved and understand the purpose behind some of the issues. Communication is extremely important between the volunteer groups so that they understand the strategies and can align their activities with the overall board strategy. The decisions that are made in the boardroom should be communicated. The board has to make sure that there is a communication plan in place to let the different committees, volunteers, and members know what is going on. Transparency is a main component of this.

Board members should always be looking out for key volunteers, people interested in watching the organization grow and evolve. The board is not going to select all of the volunteers and it is not about hand-picking the leaders; however, it is about providing the environment to challenge the leaders in the community and set the standards for what it takes to be a leader. The board of directors should be identifying and developing volunteers into leaders of the community. If the environment does not permit users to participate in leadership and growth opportunities, then the board is not fostering the needed culture and succession planning will become very difficult.The board of directors should be setting up the qualifications and what is expected to be part of the leadership. This will prepare volunteers for leadership roles and make succession planning easier.

All of these pieces form the governance framework. As a board member, strategy and defining a clear functioning governance is going to be one of the more difficult things that need to be accomplished, but this sets the baseline for everything that the board does, and it provides details about the volunteers and committees.

Being a board member is extremely challenging, not only in terms of time and commitment but also in creating frameworks and governance. The board needs to first understand the culture of the group and discover the areas that are of importance. It is their responsibility to lead the community, and validate the priorities and alignment of the activities with the overall vision.

The board has a great responsibility with these activities. It also focuses on the strategic aspects and not the individual tasks. The board needs to lead, plan, and enable the community to stay true to the core values and purpose for the user group.

Table 2-1 gives an example of potential roles for a board of directors.

Table 2-1. *Board of Director Roles.*

Role	Description
President	Represents the board of directors and community. Provides guidance in strategy and planning and leads the board meetings
Vice President	Supporting voice for the President and board of directors. Trusted council for executive decisions for the board. Could also hold a roles in membership, marketing or other committee
Director of Finance	Works on budgets and reports to board the financial plan and actuals
Director of Membership	Develops communication plan to members, new membership programs and outreach to the community. Lead committee of volunteers to work on these programs for members
Director of Partners	Works with the vendor with the membership in mind. Provides feedback from the community, and communication back from the partners. Develops outreach to other partner vendors for involvement and support in programs
Director of Marketing	Plans marketing messages, setup social media communications and creates marketing strategies
Director of Special Interest or Other	Responsible for special areas, might even be special interest groups and committees. Other is just a placeholder for potential opportunities

Committees

Committees are organized groups of leaders who set direction and plan strategies. However, unlike board members, they spend time completing tasks.

There should be a committee chairman. The committee would be formed to start to organize the group and plan activities. The committee should also be tasked with the development of the framework for a future board of directors. The framework should include how the committee is going to design the strategy for the user group. The committee members are going to be the ones also getting the work done here in all of these areas.

Committee responsibilities could include education, membership, and other special areas that need focus for the user group base. Committees are an excellent opportunity for volunteers to get involve and develop leadership skills as they start to get more information from the board of directors and steer a group to accomplish different objectives.

In order to keep the committees aligned with overall board strategy, it is recommended that a board member or a liaison is part of the committee. This will engage the committee in the strategy and feel responsible for their contributions to the goals. Even with a specific focus area, not the complete user group leadership, the committee should be organized with a leader and a team ready to roll up their sleeves and ready to help.

For example, the committee might be planning a webinar as part of the education strategy. The committee would be scheduling the webinar, setting up the speaker, scheduling the marketing emails and setting a plan to follow up after the webinar. All of these steps can plug into the different areas of the group such as education, marketing and communications. Committees can be executing the plans or finding the volunteers to help out. Each of these tasks will help to organize volunteers which we will discuss in the next chapter.

In starting a group it is goingIt is important to take time to put together the different committees. The point of the user group is to be for the users, and the committees can have the opportunities for the users to learn not only the leadership skills but about the issues they are facing at work. An education or conference committee might be the most popular because there is some glamour with the work that needs to get done.

I started on a committee for a special interest group. I found out that there was a resource with the vendor that was interested in talking with users about the needed cases, and I was able to get more information. The effort here really helped in not only the company and the ability to provide a solution, but provide me a real look into the value with user groups. If I could do this, do a little outreach and come up with a great solution, someone else might be looking for the same type of answer. I was hooked and wanting to make sure that there was a way for others to find out too. Continuing to gather new requirements, the committee was built to provide communication between the vendor and members. Webinar schedules were developed and the special interest group was off and running. Not only was itIt was fun to work with the different groups. I immediately experienced the benefits of solutions for my day to day tasks, and looking back started to develop these very important soft skills.

Committees are an excellent way to provide volunteers with meaningful activities to support the group and express their ideas. Not only will these close relationships between committees and the board of directors provide a cohesive strategy for the group and help to get things accomplished but it will also build leaders.

13

Leaders

One would think that any of the board of directors or people on the committees would be in the group leaders, and with the titles that might be case visually to everyone. However, the volunteers and members are important leaders. There are opportunities here to grow as leaders, develop skills in participating in the user groups. For professionals, development of leadership skills are important not only to be successful but also for career satisfaction. Continued growth and learning connect people with their careers and community, and why not learn skills that will allow you to enable others to do the same. Some people have natural skills in this area; others have to work harder to gain this skill set. Even technology professionals becoming leaders in their field would need to gain knowledge about themselves, working and guiding others to complete projects, meet objectives and work with the business.

Maria Anderson, president of the Independent Oracle User Group, stated in a recent article in Database Trends and Applications, *Leadership and IT – Closing the Gap*, "Leadership is about developing communication skills, being genuinely curious about your organization's business, and being creative about how you can add value." Whether it is creating a connection between the business and the technologies, vendors and users, or looking to bring teams together to be successful in projects, we can all play a leadership role and contribute with our thoughts and ideas.

Leaders need to be in the different roles for committees and board of director positions. The governance framework is a core artifact for the user group, which explains the different leadership positions and their definitions that are needed for the user group to function. To include in the framework is the definition of the role for president and other members of the board of directors and structures of the committees which include a director as part of the committees. Roles can change over time so the descriptions should be kept at a higher level.Leaders should be sought after. It is not the ideal situation to wait and see who shows up; actively engage the user community in this area of the framework, communicating the roles and opportunities and grooming community members for these roles.

Succession Planning

Investing in people in the community and at the workplace provides them with the skills that are needed to go beyond their current role. This might sound easier than it really is—a little bit of training here, some mentoring there—but it requires a well-thought-out plan. Succession planning needs to be included in the culture and governance framework of the user group.

As much as we think we have endless hours and ideas, bringing in others into the mix of the leadership is going to ensure that the group will still be going strong even after having to step away. I will admit it even with the difficulty, this is such a rewarding task and should not be taken lightly. It is incredibly wonderful to look back at a group and know that you influenced change or development in another person to stretch beyond what they thought they could do. It should always be a goal to have the group do more afterward and continue without an individual having to be the one driving it and pushing for all of the activity and involvement.

Not everyone is going to be at the right place at the right time when it comes to leadership commitments. When the board needs people or a committee needs to be formed, those good leaders might be sparse and extremely hard to find. With a user group, this is an activity outside of the regular day to day, and we all have life commitments and priorities. It is finding the right people and preparing them for any time and then most importantly, teaching them how to find people to do the same.

If we look at governance framework, there would be a place to have leaders of committees should be considered for upcoming board positions. Or a governing council on the board will include the board members that are being prepared for the leadership roles in the board. It doesn't always have to follow the exact plan, because people get hired by the vendor or something can come up in a different aspect of their lives, but taking these risks into consideration, plans can been formalized as part of the governing board and committee members.

Table 2-2 provides just a few examples of what to include as part of the framework, and something concrete like this does help to talk to members and level set expectations. In starting a group, the qualifications might not yet be possible, but focusing on some of the people that have been working with the vendor for a while and understanding the culture of those that work in this area would be able to add value to the leadership team.

Table 2-2. *Governance Framework—Qualifications*

Board of Directors	Qualifications
Involvement	Should be involved in a committee for two or more years.
Development	Continued interest in growing and developing not only in focus areas, but in leadership
Roles	Demonstrate leadership skills based on role in: Company, user group, other outside activities
Committee Member	**Qualifications**
Involvement	Should be a user group member for at least one year
Development	Continued interest in growing and developing not only in focus areas, but in leadership
Roles	Brings skillset to committee either experience or knowledge that can be leveraged. Working with the vendor for at least one year.

As one moves through the volunteer aspects of the group, you should be always looking for individuals to encourage to get involved. Attending a conference and learning is one thing, but giving a presentation you actually learn even more than sitting in the audience. Speaking on a topic is great, but then reading through hundreds of papers and seeing what is important to others provides a bigger picture as the papers get selected for the next conference. How about leading the conference to also get involved in some of the planning to make it a better user experience and that would bring you right to a board position before you know it because you now understand what some of the members are looking for, how to engage the volunteers and have a stake the game in talking with the vendor for enhancements and other support areas that you have been able to deliver back to your business.

Discussions at the different events, volunteering starts these conversations, but having something that is a tangible way to start is going to provide them with an immediate action item and show almost instant gratification. The reason why the leaders are enthusiastic is that they have already experienced the value and know the very practical ways that it is helped them in their careers. It is easy to explain and demonstrate. Pretty soon people will be rolling up their sleeves and asking what can they do. They enjoyed seeing that piece as part of the user group, how do they get involved. The framework with the steps and qualifications to be part of it will give them a road map on how to get as far as they would like to go, but there are practical ways to engage them:

- Organize a webinar

- Present a webinar

- Write an article

- Write a blog

- Participate in a membership focus group

- Participate in beta testing

- Gather information for an interest area

- Send out a marketing email

- Plan an event

- Contact the vendor on a specific topic

These are just a few ideas for immediate action. This is not giving away board secrets or discussing items that was intended just for the boardroom, but maybe a goal of a committee or a monthly plan for the webinars. This allows for a different perspective and view. These are the soft skills that get developed in understanding the group and how to participate and lead where you are at in the group.

Mentorship should begin in committees, but as a board of directors, they should be training their next board member. As soon as a group is formed to work together, they have to be planning to explain the culture, develop leadership skills in the volunteers, and demonstrating what they have learned over the years. It is amazing how much can be learned in a short period of time and not just with the vendor that you are working with in the knowledge gleamed from working with them on the board, but how to communicate and plan. The board liaison on the committees is important for keeping the strategies aligned with the board but it has the secondary function of mentoring the volunteers. Remember the goal is not to replace yourself with yourself, you are unique in what you bring, but have a succession planning that will include people to continue with the culture and understanding the governance of the board. Having different ideas and a place to discuss them is going to be important for hashing through the strategies, because all of the same ideas will get you the same answer, and especially in technology, you are either adapting or falling behind. Mentoring will include setting goals of where they would like to be with their involvement in the user community, assessing areas that they need to develop in to be effective and helping them to understand the culture of the community to be able to support it even better than before.

Summary

User group governance includes quite a few different pieces that is very valuable for the group to grow and evolve into a community that is effective and enjoying what they do. Governance framework includes how a board of directors is setup, defines the culture of the group, supports the group with committees and plans for succession.

Starting a user group gives you the advantage at starting with a solid framework that can be used for governance and development of the culture instead of incorporating it afterwards. Leaders of the group have opportunities to grow and develop leadership skills as they work with the community. Providing mentorship as a part of leadership development is a perfect way to lead into succession planning and train the future leaders of the group. The exciting part of creating the framework and strategies, it is easy to communicate to others, so the overall mission of the group can be passed on, at the same time with the different leaders coming in, it can be made stronger and even more valuable for the community.

CHAPTER 3

■ ■ ■

Building Leaders Volunteers

Volunteer engagement is instrumental in developing a collaborative and vibrant community. The designated leaders on the board have strategies and committees in place to accomplish the many tasks to sustain the user group, but it is with the strength and energy of the volunteers that the community will be more than just maintained. The user group will provide more than just education regarding technologies and products as well as opportunities for personal development and growth. In technology fields, users might be more focused on implementing and engineering solutions and do not have as much time to develop leadership skills. Even with technology focus there are ways to take initiative to solve problems and mentor others leading to soft skill development. The user group can provide a place to cultivate leadership skills and then be able to demonstrate them in volunteer positions or in the day-to-day work environment. Even without the management title or specific managing responsibilities everyone can be a leader and can improve leadership skills. User groups are an ideal place to advance , continue to learn, and to prepare volunteers for leadership and other roles.

User groups are for sharing ideas and experiences so that you do not have to be isolated and do it alone. There are more examples and cases seen in the group than you would see in one company. The network available through the user group is instrumental for understanding different views and situations.Most people do not like going to the movies alone but probably have gone at least once. Why would this happen? Because the movie was something that we wanted to see, and maybe there wasn't anyone else around who was interested in the same movie. Some people might prefer to go to a movie alone. However, after seeing the movie alone and being excited about a scene or thinking about how that technology dashboard in one of the superhero movies would be really cool if you could have it in real-life. What about that security hack, it could never happen in a company. These topics leave us wanting to discuss with others and find out what they are thinking. Movies might be more fun to discuss than work-related items, but those involved in user groups feel they have found their calling, and it is something that they can always talk about and be excited about. This is how involvement starts, realizing that in the user group there are others extremely excited and enthusiastic about what they do. They have a passion, and that passion is even fueled by teaching others, and sharing what they have experienced.It is exhilarating to talk to others and solve a problem or a challenge that would not have happened without additional input.

Once we have recognized that the user groups is a place to get information as well as increase our excitement and work with others to share, this will allow us to build up the volunteer base and fuel the passion that the group has about the mission and goals of the user group.

Volunteer Activities

The education or hearing about a use case might have attracted a person to the group in the first place. They see the people speaking and those on the board meeting people, talking and planning, and cannot really imagine themselves doing anything for the group. They are worrying about any time commitment, see some of the benefit but have not yet experienced it for themselves yet. Because of this, activities for volunteers need to be planned in appropriate sizes and areas to be able to include several different types of users.

This means planning small tasks to get volunteers started. Tasks need to be something specific, because it gives the responsibility and ownership to the volunteer. In just saying we need volunteers, for what? Having the activities planned out to say we need volunteers to run a webinar, or we need volunteers to read ten articles to provide comments on. These are very specific things and someone hearing that, reading ten articles about stuff I want to learn about, I can do that. Or run a webinar might have a couple more questions, but in stating need someone to feed questions to the speaker or ask a question if no one chimes in. Again, the activity is the right size and specific for someone to think that is an easy step. Asking a volunteer to immediately jump in and speak at several webinars or run monthly meetings for a regional group, might be too much at one time. Other volunteers might be ready to take on whatever you can give them, and groups of activities can be setup for these different volunteers. Appropriate size of a small might be something that can take a volunteer less than 30 minutes to do. Then there might be regular activities to introduce speakers on the webinars, or help send out messages on social media or be part of a twitter chat. The next chapter talks about some of the activities with social media, and it that should give you a few more ideas for tasks for volunteers.

There might be people in the user group that have great experience and knowledge, and it would be ideal to have them volunteer to speak at an event or even webinar. They might not feel comfortable speaking, which could be a great development and building opportunity, but they could also provide online content, either in the form of a blog or an article. Other ways might be partnering volunteers to work on a presentation together, builds confidence, encourage volunteers and as we will see later develops leadership too.

Some more examples of some volunteer tasks could be reading through papers and presentations for gathering the tips and tricks, contacting speakers to confirm participation in conferences or webinars. The best part about these easy tasks is that the knowledge that is gained by the volunteer. They start learning and wanting more information and want to read more. There might also be a very popular speaker that the group really respects and the volunteer has the opportunity to either meet the speaker directly or virtually. As conversations like this start to happen, there are chances to ask questions. What a great volunteer bonus to get a bunch of research done and one on one time to ask a couple of questions of these superstar speakers. Figure 3-1 shows examples of activities with the smaller ovals being those with less time commitment. This is also not meant to be an exhaustive list.

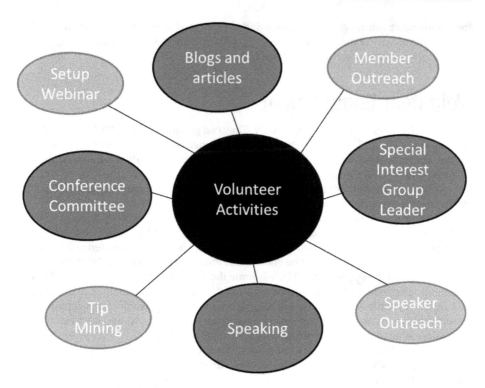

Figure 3-1. Examples of volunteer activities

Another activity might be providing feedback on topics of interest or what they feel are hot topics and areas that continue to be important for all members of the community. This might not seem as important, but if you consider the current projects the volunteer might have, they might be getting extra help in this area or be able to be in contact with the vendor on specific information that will help them with the project. They might influence a track for the conference on the topic, but chances are if there are a few projects out there with this focus there are many more, so this would benefit more of the community too. This seems like a good way to engage volunteers, to be discussed more in the next section of this chapter, but it highlights the activities that they can join in and be a part of for the community.

There are other volunteer activities that are to be expected from a user group, and these require more of a time commitment. Speakers and bloggers are going to be the experts on the products and vendor and are willing to share their ideas. Presentations and blogs take quite a bit of preparation, and without these volunteers user content would be at a minimum and there would need to be more reliance on the vendor. Conference volunteers help plan the content, select the speakers and schedule the tracks to provide the best experience for the users. Special interest groups that have a specific focus area need to be organized, and volunteers can help lead the special interest groups. Specific activities for special interests groups would be designing the group based on a charter, recruiting other volunteers for the group and planning webinars. Webinars are normally more of the activity of the special interest group because this is more of an online community but has presence at the conferences and other events.

Volunteer Engagement

Finding volunteers sometimes can be easy. They can get involved quickly because of other users who they know or if they have already experienced some benefits of the group. There are rewards that can be granted to volunteers, and other volunteers will see the leadership of the group being a way to advance in their careers.

Volunteer engagement takes other volunteers. If the group is just beginning, engaging volunteers might be the most difficult part. One of the leaders of the Independent Oracle User Group (IOUG) recruited me as volunteer. They knew of an area they thought I would do well in and asked me if I was interested. It is difficult to say to an expert that personally asks you to be involved. There is also the confidence that was given with stating that you would be good in this area or expressing that they think you would be good for a specific role. There is some thought that goes with it, and need I say planning again to learn how to engage people. It didn't take much for them to draft me for the list of people who would enjoy helping out in any way possible. I was getting a chance to learn more, having conversations with the group's experts and it was learning to manage time to make this a priority because of the benefits.

Additional strategies can be used to motivate and encourage users to get involved. Sometimes it is offering conference registration, drinks, coffee, food, access to the vendor, and marketing opportunities. It is important to make it a position that others would want to enjoy being in.

The fact that I was encouraged to participate more in the group, I am looking to encourage others as well. It is great to be part of the community that has the same interest, then to be able to contribute to the group, to quote another leader in the Oracle community, "it is awesome".

I like to use Ray as an example of a volunteer that every group needs. I met him at the user group conference, and now he introduces me to others as well. He had excitement about being part of the group, and when you would ask him to do things he would.

I was looking for a good member to help as a volunteer for pulling together content for an up and coming topic. Ray wasn't as familiar with the topic, but was willing to learn more. So he became important in pulling together speakers for the track. It was very successful, and he recruited more volunteers from those speakers to be focused on a special interest group. The next task was to investigate other special interest groups that Ray was able to help out with, and he didn't just say yes because he didn't know how to say no; he was able to learn by participating in anything that he could. This also set Ray up to be to be more engaged in conference committee. Then there was a need for someone to take over the user group journal, and whose name came to mind right away: Ray's. He was able to get writers excited about the journal, and not only encouraged them to write but also managed the volunteers properly to meet publication deadlines and communicate what was needed. He started to develop strategies around the journal and look at ways that benefits for the volunteers would help them with management of them. He even trained his replacement for the journal, so that he could find another opportunity.

It was easy, when looking for another board member, to nominate Ray for the position. It was fantastic to watch Ray grow with every new opportunity. He brought passion and excitement to everything that he did. From his position, he was grateful for each new chance and working with others. It was amazing to see him speak and mentor the new speakers getting involved in the conference. I am absolutely confident he will be representing the membership on the board, and bringing new ideas and thoughts to participate on the board. In his role, he is working with others to make the user group stronger, and getting others excited about participating too. I can't take credit for his willingness to learn and wanting to be involved, but I like to think that we can each find something for a volunteer to do to be encouraged and valued by the group. I can also hope that this example will inspire others to seek out the members of the community to participate.

Now for the funny part; something that I really like about this story and example. When I rolled off the user group board, I was recruited by Ray to participate on a committee. I am not sure he was worried about me not feeling as connected with the group or didn't want me disappearing from the user group scene, but to include me again in an activity was very special. It was his turn and he even recruited me for a couple of panel sessions at the IOUG–Collaborate Conference.

Leadership Strategies

It is important to understand that even when not in a leadership position, leadership skills help in day-to-day communications and interaction with team on how work gets accomplished. Developing a strategy to develop leadership skills for members is a great start. This will continue to build a base of leaders.

Figure 3-2 shows the progression from member to board of directors.

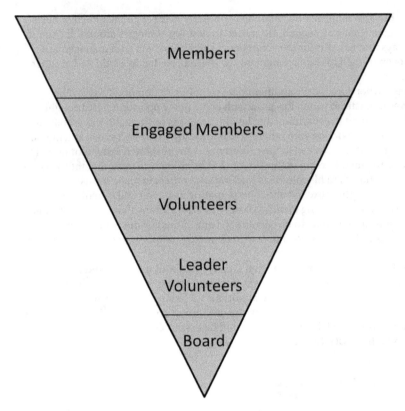

Figure 3-2. *From member to leader*

This upside-down pyramid demonstrates why members are at the top: this figure shows a filter. There are going to be more members than the board of directors, which makes perfect sense. As members grow in their understanding of the user group and benefit of volunteering, they move through the filter. The members are also at the top because the members are the community to be supported by a strong base of volunteers and leaders. Leadership is not about being on the top. It is about serving others and working for the benefit of the group. The leadership strategy is to demonstrate and teach the leadership skills to cultivate the volunteers. As the members become engaged in the group and start to participate in more events and discussion. Members that are engaged already provide an overall benefit to the user group by attending and participating, and it shows the interest areas and what topics are important. Engaged members will filter down into volunteers. They will start to see opportunities where they can participate, and as they learn the leadership skills and start to participate in the communities' leadership strategies. Even if engaged members do not move into volunteer roles, they can still learn the planned leadership skills to help them in their companies and with their careers.

Technologists normally have their focus on the technology, code, and learning new technologies and versions. Developing the leadership skills in the group can get you noticed in your organization, not only as the expert in the technology but also as someone with the skills to provide leadership to the team and proper direction. Critical decisions can be difficult, but as these skills are developed as part of the technology learning, these decisions get easier and help with stressful work situations and architecture decisions as well.

Leadership can be learned and it is a skill that can be practiced. Regardless of the role, leadership can be practiced by anyone and develop the consistent behaviors. The user group can provide opportunities to participate in volunteer activities to continue to practice. As leaders are cultivated as part of the strategy, the user group will grow just because of the engaged members and volunteers.

Leader Volunteers

Learning how to create and nurture collaborative relationships with others is an essential leadership skill. This can be done with other volunteers and members, and brought back to you day-to-day job. The importance of collaborative relationships seems to be understated in most cases. There are several groups in an organization that rely on other groups. There is so much potential to be gained by building and developing these relationships. Building collaborative relationships involves reciprocity or helping others. The teams that appear to be working in a silo are going to be the most difficult to reach out to but it just takes starting the conversation. The benefit to the organizations is going to be incredible if the relationships are built and continue to be maintained.

Motivation to work as a leader and collaborate with others in a company comes from producing a successful project, product or service. There is also the pay factor, to earn money, for a job well done. Money is not a motivation that can be used with volunteers and in a user group. The product of the group might just be the overall strategy and goals, but there are not always projects. Many technology professionals must want to learn and grow. The continued learning is part of the ever advancing technology field. Once we stop learning, our skills become obsolete. To be able to be self-directed in these areas help for motivation to participate more in the user group and develop into the leader volunteer. The mind-set of the volunteer is to pursue a vision and know that there are others wanting to do the same. From the board of directors standpoint it is imperative to share the overall goals, vision, and strategies with volunteers to help them stay engaged and motivated.

Volunteers have to want to collaborate, even if it means reaching out to others that you may not know. It will involve stepping out of a comfort zone to continue to learn and grow. It means helping others to do the same and wanting to help others and not for the sole purpose of your own personal gain and career advancement. The motivation to be in a volunteer role will come from helping others succeed; this will lead to authentic long-lasting relationships. If you think about how you would like others to collaborate with you and build relationships with you, you can turn that around to create the new relationship and your authenticity will be very apparent. As leader volunteers, the interest in helping others solve problems and find solutions will show through and make strong relationships.

I found myself in a job that was very task-oriented. In technology there is nothing wrong with staying on task to provide stable, consistent environments, and not all technology professionals are going to want to pursue more outside of completing their checklists and task oriented activities. However, I had felt that there were talents not being utilized, areas that I wanted to grow into and career goals I had set. My current position and career plans were a big driver behind me looking for other activities to cultivate the leadership skills. This is what was driving me to learn and adapt by finding something outside of work with the user group to advance and start using some of these skills. Not only was I pursuing the new technology information and learning about the different skills, but I was really becoming a leader. It was the motivation to collaborate, help others, and use my talents for the vision of the community that help do what needed to be done and continue to work in planning for the user group.

Leader volunteers have a drive to learn, adapt, and advance. They set goals based on the strategies of the user group and are willing to feel uncomfortable and push through the issues. As the work is accomplished, the goal is reset and evaluated to see what is next. Possessing adaptability helps advance your knowledge and reframe these experiences as opportunities to learn.

Leader volunteers develop the following skills:

- Collaborative relationships

- Motivation coming from achieving goals

- Ability to reset goals and reframe experiences

- Adaptability to different experiences and situations

- Opportunities to bring others along

- Constant drive to learn

User groups should be providing leader volunteers the opportunity to grow and practice the skills that they learn. The roles and different committees provide the leader volunteers a way to participate and develop. Leaders are only growing if they are learning. There is always something new to discover about you, new ways to adapt, and new relationships to cultivate.

Working toward a personal vision doesn't support the user community. As a leader volunteer you want to work toward the vision of the group and continued learning and adapting is key here. Think beyond the technology and the basic skills and try to understand how these skills and technology can advance your business, department, or team. Become a leader volunteer by understanding the vision for the user group organization and bring others along. Helping others in the user community is really what a user group is about.

Leader volunteers need a place to practice their leadership skills and it needs to be part of the user group culture. These volunteers are going to be the ones to get members engaged, excited and bring them along for growth and development in the user community.

Summary

The leader volunteer is going to be passing along enthusiasm to other volunteers and the members. Understanding the opportunity that is out there to be part of the user community and appreciate the benefits to be able to continue to learn and collaborate with other teams. Leadership development skills are not just for the potential volunteers but provide a valuable skill set for the community to learn. Everything from communication to building collaboration relationships with workplace partners, teams and community members.

There opportunities to participate in user group activities and get more involved. Leader volunteers comprehend the strategies and visons of the user group and participate in the activities such as leading special interest groups, writing a blog, writing an article, or speaking at an event. These activities all find ways to help continue the growth and practice leadership skills. The initial planning of the group, governance, culture, and vision provide what the group needs to support volunteers and start engaging the volunteers for more involvement. Beginning groups will need to work harder and smarter to engage key volunteers and communication will be very important as the group is developing.

I look back at the engagement areas as I was pulled, maybe gently pushed, into in order to get out of my comfort zone, and how the collaboration between the other members and leaders, really made a difference. Being encouraged by leaders or others in the community increased my motivation to wanting to continue to learn and develop these skills. Because it is others in the community and volunteers that is encouraging the involvement, I wanted to also give back, and just like my example with an awesome volunteer Ray, I was excited about opportunities to bring others along too. Watching the different members grow into volunteers and then grasping the concept of leader volunteers as the leadership skills improve.

It really is incredible what a volunteer could do in building the user group community. Engaged and leader volunteers are what makes up the community. Planning to utilize the different skill sets and levels of engagement allow different strategies and activities to be used. Education around the product and vendor is definitely important for the day to day knowledge that is needed, but to become an expert or demonstrate expertise in these areas, the soft skills of leadership, communication and collaboration are what is needed. The user group needs to have a strategy to educate members around these skills and provide areas to practice and to cultivate these skills.

The user group purpose, strategy, governance have been discussed in previous chapters. With this chapter we talked about the volunteers that are needed and their engagement. Next we will look at the planning of the events and other activities that the user group can do based on this initial setup and getting the group started as the user group and community are starting to take shape with others being involved and promoting and growing the group.

CHAPTER 4

■ ■ ■

User Group Planning

A user group is to provide activities for the members, including education, networking, and relevant resources. Event and activities can be planned to get the group started. If the group is already established, there can be additional planning to make sure that the events are what the group needs.

Planning includes examining different options: Does the vendor already have an event? Does the group prefer in-person or virtual activities? But before all of this, there is preliminary planning to be done.

Board strategy needs to be taken into consideration when making plans. Careful planning of activities is necessary to not overwhelm the community and still provide specialize areas to benefit the members. Too many choices might dilute the focus areas and not highlight the key topics. Especially in technology areas, users might be left with several different areas of importance while still supporting legacy environments and needing constant information to manage both sides. As volunteers, we have to understand this and with our resources balance the new, popular, and support topics.

Planning includes meetings and activities but also includes interaction with vendors, any publications, blogging, and other media. The demand for this information is going to come out of the community and user base, and it is not easy putting together a strategy that is going to provide a large group of users everything they want or need. Focus groups can help determine the most common needs and delivery forms. User group surveys can also help out to get details depending on how new the group is. Working with sponsors is going to allow more of these events to take place, but the benefits need to be part of the experience for the sponsors and users. There are great opportunities in putting together a plan for the users to get the most out of the community. The more ways to engage the users, the more ways to be involved and support others in the community. The following are some great ways to engage users, vendors, and sponsors, and purpose of having these activities as offerings.

Events and Meetings

Events and meetings are obvious ways to be part of user group activities. Vendors are likely to have their own events or conferences. Resources will play a big part of what type of meetings and how many but there are ways to reach a goal with additional efforts with sponsors.

Event planning can be easier when an association management company is engaged. This management company can leverage their connections and negotiation skills to keep costs in line. Having staff to dedicate to a conference will free up the volunteers to confirm the content and speakers. Experience in running conferences and events is beneficial so that small items do not get forgotten, and so that attendees have a great experience. When deciding on which association management to use, speak with other user groups. The staff becomes an extension of the volunteers to help run successful events and conferences.

When planning events and meetings, it is important to look at year-round activity. Conferences can be a focus point with other meetings supplementing throughout the rest of the years. Ideas for events can include:

- Conferences

- Workshops

- Social Events

- Lunch and Learn

- Meet-Ups

- Virtual Events

- Webinars

Conferences

The user group can work with the vendor to participate with speakers or a booth to allow introductions to other users that might not know about the community. User perspective is important for the vendor, but it is also what makes the user group conference important. It is because of the insight of either how to use the tools and products from the customer side that provides this valuable information. There might be other competing events or conferences and in order to have any type of user group conference, these things must be taken into consideration. What are the opportunities for the members? How is the conference going to distinguish itself from the other events? Is the user group more dependent on the vendor for content and information or are there experts in the community to allow the event to be more independent from the vendor?

A conference is a great opportunity to network and hear from experts in the community. The schedule for the speakers should take into consideration the new topics and current ones that are relevant for the community. In deciding on the topics, it is necessary to approach as well as have an open call for speakers. There should be a few topic areas that are going to be the main focus, and the conference can be organized around these topics. This makes it easier to find the right speakers, and the attendees will be able to grab the right sessions.

Educational content of a conference is extremely important. This is the main reason for the attendees to be at the conference. Education needs to be organized to show its value, and then there needs to be other conference sessions or time to give opportunity to be able to ask the questions and gather the information from the other experienced users. The attending experience is important, but the value comes out of organizing a great educational event.

Conferences work by planning the big stuff first; education, the main networking event, and the schedule, and the rest of the little details can fit into this bigger plan. The schedule would be deciding how many days the conference should be, and how many of the hours are education versus networking and vendor events. The space and the hours of education sessions give you the idea of how many actually sessions you can have. Fuller sessions are going to make it more enjoyable for the speakers and overall feeling of the conference.

Setting the topics as early as possible for the conference allows the speakers to address needs and topic areas. Grouping the sessions so that conference attendees can either focus on this specific area or look at other sessions to supplement will give them the education that they are wanting. It also will give justification to companies for attending the user group conference instead of the vendor conference.

Workshops

Smaller events such as workshops can be used to provide one- or two-day sessions outside of the conference. Maybe workshops can even be done every quarter, depending on costs and speaker resources. Regional workshops might even be held in different areas than the conferences to give those the opportunity to attend that might not have the travel budget but can attend a regional event.

The database professional space might have conference topics around database manageability, development, business intelligence, database security, new features, and high availability. These topic areas are very difficult to cover in a day or even two day event. But a workshop might take the same speakers under one or two topic areas and host them in a workshop. A one-day workshop might take the subject area of database high availability and have a workshop with sessions and maybe even include a hands-on part of that to round out the experience. This type of setup works well between conferences and provides additional education to different regions. It should be part of the year round strategy.

Social Events

Social events are a part of conferences that allow for networking opportunities. Meeting other users to talk through issues, problems, and successes is a main reason for the user group in the first place. The conference either the user group or vendor conference can have a couple of social events to encourage these connections and interactions.

However, social events can stand on their own, too. They can be a focus of a regional group to meet people in the region with the same experiences to provide educational content and networking. A social event regionally might have one speaker take for an hour or so, then have some sort of food, drink, and further conversations. This would also be a great event to be recruiting volunteers for other activities, which is discussed in another chapter, such as speaking or helping out with another committee.

Social events are normally very simple and are perfect for vendors to sponsor the food or host the meeting. The cost to the user group should be minimal, which again makes it perfect to have as an event between other type of events to have year-round education and participation in the user group.

The social events can be setup to a regular meeting, maybe once a quarter or something along those lines. Regular meetings that are held do allow people who can't join for one event be able to plan to be there for another event.

Lunch and Learn

Lunch and learn is similar to a social event but is more simple. A regular restaurant or meeting place can be established and people can bring their own lunch or have a vendor sponsored lunch. There might be a speaker or a round table type of discussion. The topic can be sent out with a reminder email and format. The bonus would be to have a vendor sponsored lunch where they have a few minutes to address those who come and then discussions can be around that topic.

These type of meetings are going to be more local and not really attract too many from other regions. They are probably going to be smaller, but give the users a way to follow up with each other on discussions that might have come out of presentations, vendor announcements or conferences. Lunch and learns should have a way to present on a topic for a few minutes and give time for discussion. There isn't much planning that must occur here, just simply having a place to meet and setup a discussion for the day. The frequency of these meetings can be once or month or quarter. It would fit into more user's timeframe in that they can take a lunch break. Additional planning would be needed if the vendor sponsorship would be included to make sure that there would be more users attending through additional marketing too.

The other advantage of a meeting such as this, is that it can easily be run by a few regional volunteers and give them an opportunity to pull together topics that make sense to them to cover and support. If the board strategy doesn't have a built in process to manage these type of groups, it should be considered as part the framework and culture of the group.

Meet-Ups

Meet-ups are going to be almost like Lunch and Learn, maybe just without food. These are just in person events that will allow the group to meet and discuss current issues. The meet-ups I have been a part of were just to have straight talk about the vendor and experiences. Everyone brought their own coffee or food with the main focus being around the discussion. This again can be a more frequent type of event and should be scheduled to allow of a reoccurring schedule to be part of the calendar. Local volunteers can be used again for these events, but it would be good that they were on the overall board calendars to bring awareness and visibility into the regional groups that might be forming

Networking Opportunities

Networking should be high on the list of priorities, as part of the strategy and purpose of the user group. Networking and talking with other users might have been the basis for forming the group in the first place. Networking doesn't just mean connecting with people or users; it means having a group that can provide answers and solutions or use cases that others might not have even thought of yet.

Being able to build a network by meeting with experts at events or finding other users with the same issues provides a great reason to attend the in-person events. In technology user groups, the networking and being able to reach out to other users for ideas or confirming setup. Those part of the user group tend to have fewer support calls because of this network and trusted sources of information. Networking with other users provides more excitement about the technology, products, and even the day to day.

Networking opportunities is a main reason why I stay involved in user groups. There is something that brings people together from all over the world to be able to talk together in a familiar language. I know if I have questions about any product or area I can find someone I can talk to about it. This is especially important for the companies that do not have a complete group of people with the needed expertise, the user group expands this level of knowledge and being able to talk to others to solve a problem or confirm a best practice is extremely valuable. I have heard from users attending conferences that the conference has paid for itself just because of conversations held with the experts in the hallway, and then having an email for follow up afterwards. It is exciting to talk to other users and being able to network and hear what it is important to their projects, what type of answers they are looking for and just develop the connection.

Conferences are the ideal place to have some key networking opportunities. The conversations are going to happen in the hallway if the schedules are setup properly, but there should also be some planned special networking time. There have been different activities in the evening that have worked really well at some of the technology conferences from install-fests and hands-on lab time with pizza and drinks. Even a different take on games to include a large group of people from jeopardy to role-playing, and these give the users additional way to connect, which then spurs on even more hallway discussions the next day. Networking opportunities at in-person events are great for building the relationships to be able to contact other users after the conference and look forward to seeing the users next time. It definitely starts to build the network to look forward to each year and chances to talk with them throughout the year.

Events and conferences are not the only networking opportunities. There can be discussion forums that allow for responses and discussions. These discussions are of course not just networking opportunities, but it does get the community involved. There are ways to contact people with discussions and continue to ask questions. The following two sections about social media and virtual events present a couple of different ways to have online networking opportunities.

Networking is also an important benefit of the users and volunteers as they can work with the vendor. Product managers and possible developers of the applications provide an incredible resource to the community. The solutions and information for common problems and issues is why a network is so valuable. The users that get these types of answers are also going to be more involved in helping others, so the vendor networking here and investing in the community gets value back with others to talk about issues and solutions.

Social Media

Social media is an interesting topic. In some groups the social media component is going to be part of the main focus of what is done, and the strategy might even revolve around social media. Other groups might see social media as something they should be doing, but not have a specific plan on how to use it. Still other groups will find that social media will be able to reach out to another audience that they were not expecting and the value of marketing with social media. The way that we communicate has changed over the past couple of years and with the social media available it is going to be part of any group that is looking to connect users throughout the year.

I would consider myself a "social" tweeter. If I am attending a conference or an event, I will take the opportunity to talk about things happening at the conference. Most of my tweets come from this time either promoting a session or activities that I am enjoying at the conference. Even though I have more people and users around me at these events, this is just a fun something to do, which can get responses faster from others at the conference. This is not necessarily a bad thing, and I have found myself doing a little bit more throughout the year. This might be typical for other users in my focus area and especially when some companies do not allow access to the social media sites. Understanding the users and some of these limitations will help in developing the social media plan.

Social media is a powerful tool for user groups. Even if the culture is not heavily involved with social media, it does provide a space to drive some interaction between users at events or virtually between the conferences.Planning social media includes choosing sites, group names, and monitoring. The typical sites used are Twitter, LinkedIn, and Facebook. Social media can be used to promote different content, conferences and build interest for discussions for networking. Imagine an article that just came out that would be of interest of the users, links could be added, summary about the article and opportunities to add comments to include other user feedback. This can be done very quickly with social media and start to build up groups to have these discussions. Posts can include topics, speakers, and special events. During events, those who were not able to attend can feel connected by social media and it gives attendees the chance to talk about what is going on and what they are enjoying. Even the data coming out of the social media for the conference chat before, after and during, can provide key information about what the users are looking for in content, answers and where they had the most fun. It seems like a great tool to introduce a new idea and gather up interest for possible programs.

Another area of social media to explore is videos and postings on sites such as YouTube and blogs. If you think about all of the things you might have learned recently to do from YouTube videos such as cooking a recipe, learning an instrument, learning programming languages and even figuring out how to do something on Minecraft, you start to realize that tThis is an excellent tool for the user community. The speakers and expert sessions are even more accessible. Making a short video can provide quick information and quick videos can even be collected for tips or from conferences with additional extra details about what was learned. There are multiple ways to leverage videos and create a channel for the user group.

Blog posts can be something that the user community might already be doing, but the volunteers and leaders can be blogging about different information for the community; opportunities to get involved in the group, upcoming events and educational material. Blog posts can link to different information to lead to events, networking opportunities and other discussions. Blogs are just not as formal type of writing but should have relevant content and information. Promoting speaker blog sites can also encourage involvement in presenting and developing user connections. Leveraging existing bloggers in the community will design a way to highlight these topics and interest areas and create some discussion around them. Blogging can be included in the plan as to build out a team to regularly post information and gather topics to include as part of the user community. The blog would be focused on user group topics and aligned with the strategy of the group, and if subject areas need to change working with the volunteers would allow for this to happen. Either way the speakers and bloggers are shown as experts in the community and provide relevant content to the users.

Social media is not just a tweet or a post on linked-in, there are plenty of opportunities to use these tools to link together a complete strategy from marketing events to building networking opportunities for the users. Planning for social media should include activities that bring different events together and show the value that is being brought to the community. The marketing here might even be reaching users outside of the regular community members, and providing discussions to open up the networking opportunities.

Virtual Meetings

In-person events have speakers, a chance to ask questions, and hallway discussions. There are also activities that can be held all throughout the day and evening socializing events with food and drink. Virtual events have speakers, a chance to ask questions, and a possible way to have virtual discussions. Virtual offerings can be added to in-person events to take a session track or topic area and have the sessions be added to the virtual content live at the same time of the conference. With software this is possible to hear the speaker, see the presentation, and have a live Q&A session. These extra packages do not normal take away from people attending the conference but provide a way for coworkers to also attend virtually or those without travel budgets to attend. A virtual event provides an additional benefit to the members and can supplement the budget for the conference, too, which is very helpful in the beginning of the user group.

Technology groups even appreciate in-person events, maybe because it is a way to get away from the desk and some additional distractions to be able to learn, but to round out the education all year or to be able to allow the community to meet more often, virtual events and meetings are good ways to cost effectively meet. There are several ways to host virtual events. It can be a one- (or more) day event, maybe just morning or afternoon depending on time zones for attendees, with new speakers or sessions from a recently completed conference. Virtual might even mean a series of webinars. If you are looking at year round touch points a combination of these virtual events could add content and value for the users. Table 4-1 describes a possible schedule for virtual events.

Table 4-1. *Virtual Event Schedule*

Event	Schedule	Topic/Format
Webinar	January	Series 1 of 5—two-hour sessions with user speaker
Webinar	February	Series 2 of 5—two-hour sessions with vendor and speaker
Webinar	March	Series 3 of 5—two-hour session User speaker Case study
Virtual Event	April	Half-day event—four sessions, one hour each, combination user and vendor sessions
Webinar	May	Series 4 of 5—panel session for questions and follow-up of virtual event
Webinar	June	5 of 5—two-hour wrap-up of topic with user speaker

The plan in Table 4-1 doesn't give a complete list of ideas, but it does show an example of the different formats and possibilities for types of schedules. Virtual events can be very flexible and if a series it can be presenting something every month or even once a week. Some of the scheduling and types of webinars or virtual events will depend on the community with the openness to attending the webinar.

The leadership of the user group should look at different types of events including length and times of the events. Cost of events can be off-set with sponsorship from vendors. Virtual events does provide the education and some networking opportunities and should be considered if part of the strategy of the group.There are different opportunities for virtual events, webinars, and online networking. Mapping these types of activities to the user group overall strategy and planning these benefits bases on the community needs will help develop the user group.

Summary

The direction for the user group was set and now the planning begins. The planning for the activities, education and networking that match up with the culture of the community and the strategies. User group activities are to support the community and deliver the benefits both to the users and vendors involved. Planning is definitely key to have successful events and activities. Association management companies can assist in this area because they bring the expertise about the different types of things user groups are doing, but the board and leaders still know the community and have an understanding of what would be important to them.

Conferences and in person events should be planned as this provides excellent educations from users of the products and allow users to network and have the hallway discussions to troubleshoot issues and find out what others are doing to implement solutions and options. When even planning these different events it is important to look at user availability other type of events that are available to the same group and community. There are other options to start meeting if the group is new or just looking to grow out other areas. Year round activity includes planning in different areas and not just having one event.

Webinars, virtual events create activities throughout the year and allows different users to be involved. Different speakers, volunteers and types of sessions can be held to encourage participation from users and the community. Even a virtual event can promote networking and have discussions with questions.

Social media is something that can be used in any community, but because of culture of the group it might be limited. However, there are ways to promote events, and have fun with different gamification on social media. Blogs and discussions using twitter would allow different volunteers to be involved and promote different activities. They can also provide additional information and drill down into important topics with examples.

The planning that goes on for the user group can be done by the committees and volunteers of the group. The leaders and board can make sure that the plans align with the board direction and strategies. The following chapters will discuss some of the challenges in these areas and expand on how to develop volunteers and grow in leadership skills. These different areas will grow the user community and planning for growth and how the future of the group should be will allow for the best of the activities to continue and users will want to stay active and involved.

CHAPTER 5

■ ■ ■

User Group Membership

Members are the community. In forming a user group, the purpose should obviously revolve around the members. The members are a vital reason to have the group. Reasons to form a group might be a desire to have better ways to deal with a product or vendor, learn about enhancements that might be needed, or even discuss issues that might come. The members of the users group can provide help for each.

In addition to meeting others in the community, there are additional benefits available to members in a user group. Members that understand these benefits or see the value in them are going to be more engaged as part of the community. Demonstrating the value of these benefits are going to attract new members and continue to grow the group.

There are obvious reasons for growing the membership. One is to be able to represent a large diversified group and voice back to the vendor. Second is the opportunity with partner vendors that compliment this space for support and additional resources to help grow the community. Different vendor involvement can support different events and parts of conferences. The growth of the membership allows for growth of partners and provides even better feedback and resource to the main vendor and reason for having the user group.

Outreach to the community is important to let them know about the user group. But just like developing an overall strategy for the purpose of the group, membership should be considered. Membership types, options, and value are important to define. This is also an area that requires several resources and it is a decision that will set the basis for the present and future of membership for the user group. As we continue to look at the membership for the user group we will look at each of these areas in more details starting with the value that comes with membership and different types of membership.

Providing Value to Members

The benefit list has to contain items that are going to be important to the user. There are so many choices out there demanding our attention. Even with all of the resources that we have available to us at work and on the Internet, it might be challenging to provide something different or to add value. But if you think about what was just said about all of the resources, the time to go through them and weed through some of the good and bad, having a consolidated or trusted source of information is going to be extremely valuable.

Activities, information, events, networking, and education all are areas that add value to members' involvement as they join the community. There are also interactions with the vendor that will add value for the member that we will talk about later in this chapter. As you can see from these areas, they are nothing new what we already have been discussing from strategy to member and volunteer activities. There might be discounts available to different events and possibly including vendor events. Discounts have a real value that could be associated with it, but some of the other types of benefits have value they might be more subjective.

Publications are another member benefit, whether it be an ebook, newsletter, or actual paper copy of the book. With an active volunteer group it is possible to gather articles and provide outstanding details for the members. These type of publications are normally a highly rated membership value. Journals, tips and tricks, and newsletters are not only a great way to service members but also to promote the value of the user group.

Membership Types

These are the actual benefits to offer the members, but we also need to look at possible membership types, because these different levels of access. Offering a simple across the board membership might be the easiest way to manage, but there are companies that will be willing to sign up more members if they see how the company can grow from the knowledge that the community is providing. There might also be people investigating what the community is a how they can be a part of that. There are also so many free areas on the Web; if potential members are already customers, then it is possible that the vendor might be offering something as well. This is an area that needs to be researched and discover what is already out there and the normal cost to belong to these groups.

Part of the reason the discussion about the value came first before actually discussing the types of membership, it is important to show what a membership is worth. Knowing the worth doesn't mean that is the cost of a membership, but it needs to be compared and examined to determine a cost or how to promote what the group member brings and the cost from the membership. Giving membership a value even if allowing people to join for free will show some of what is needed to support the different efforts of the group. It also gives sponsors an opportunity to see what value is considered. Table 5-1 shows examples of the membership options and different levels that might be offered.

Table 5-1. *Examples of Membership Options*

Membership Type	Cost	Benefits
Individual Member	$	Individual access to user group, discount on events, newsletters, ebooks
Premium Members	$$	Individual access to user group, discount on events, newsletters, ebooks, additional education, books, or something that would add the premium value
Corporate Members	$$$	Company access to user group, discount on events, newsletters, ebooks (limited or unlimited)
Sponsor Members	??	Something included with different sponsorship levels for access to the user group and materials; might include placement or marketing materials

Table 5-1 does not include free or a minimum online options; these are also possibilities. A free membership might allow for quick startup of a user group, but it is difficult to pull back from something that was offered for free and then charge for it. Again that is where the membership value is important; perhaps it would be better to offer an initial discounted membership. There are times where joining the community without any charge makes sense. Research in this area will provide insight, but it might be that the community is not used to being charged. It might also be other resources out there that don't require anything to join. This again is a way to either show the value of the membership or fit into what is expected for the community.

Feedback from members can be obtained through membership surveys or focus groups. There is additional information that can be asked of the membership during surveys.

Feedback can be gathered in a few ways:

- Membership Surveys

- Focus Groups

- Quick Polls

- One-on-One Discussions

Membership surveys should have specific questions regarding benefits and future plans. There should be separate surveys from the ones used for conferences or other events. For this reason, membership surveys should not be done every year and should be used sparingly. Too many surveys would probably not get the full response back as needed or might limit responses on future surveys.

Focus groups can take a smaller group of the membership and get direct feedback on certain activities. Another advantage of a focus group could be to actually hold a brainstorming session to get new ideas to filter through to the board and other volunteers. Because of this type of feedback, previous activities can also be discussed with more detail of what they did or did not like and provide why, which is difficult to do in a survey.

Quick polls might be something to on the website or at the end of a webinar to gauge interest in a topic. This is something that is very short and will not require much effort on the person responding, and in an easy place that it will not take too much effort to compile results. You cannot get a lot of information this way for obvious reasons but to get a quick idea this works out well. One-on-one discussions can happen at an event or conference. This again is another great opportunity to get the valuable feedback.

The user group grows with members and the community needs to provide the value back to the members. There are limited resources to support the different activities of the community and membership dues can help cover some of these costs but people resources might need to be prioritized to meet the needs of the user group and members.

More Satisfied Customers

Members of the user group are more satisfied customers and tend to serve as references or using additional products from that vendor. How does it get that way? Is this more important to members or vendors? The reason that members of the user group are more satisfied is because they belong to the community, and volunteering gives even more of a connection to the vendor and the user group.

Vendor support of the community is important, and vendors do get benefit by having more satisfied customers. It will allow the user group community to learn directly from those who set the product direction. With user group conferences, vendor presentations provide more than just documentation and also contact information with the presentations.

Even with a good customer support, there are other products, other data, and other influences, so a user group has others in similar situations. It is not always just about the one vendor; the community then provides the insight on how it all works together. Being able to use the product to its full potential in the mixed environment is going to make members more satisfied.

It is possible that the service from vendor may or may not be the best customer service, but that doesn't matter as much if you are part of the user group. Of course, you are hoping to have a more complete experience from the vendor, but with user group support, it does make up for some of the areas that might be lacking. Because of the network, other information that is provided and troubleshooting help might be available from other avenues. Sometimes if the customer support is not what is needed, as a user community a voice can be sent back and suggestions can be made to help improve the support. Community responses and troubleshooting steps can help out and be a value part of the support. This is not saying that the user group has been formed because of the customer support, but there have been groups formed on a simple purpose to help improved this, especially if the vendor has a product that is far above or needed to meet a specific niche. User group involvement can either provide more details from the vendor or help the vendor in the support areas.

There have been vendor surveys that state the customers that are part of the user group are more satisfied. A simple thought of why this is, being part of the community, you realize you are not alone. The resources and others attempting to perform the same configuration, changes, setup provide the needed support outside of the regular vendor support.

Opportunities to Learn

Learning doesn't stop along with the opportunities to grow. There are going to be different ways to learn, and members are going to be expecting that.

Members are joining the user group for research and understanding of the product. They want to see how something is working for others and want to take advantage of that knowledge from others to gain the same experience. Since they are joining the group to learn, the membership surveys really help out what the most need is. This can provide some targeted education, maybe a series of webinars or articles.

There are a couple of things that the members start to find out when looking for growth and learning. For example, volunteering for the user group is actually a way to start learning faster and gain access to member and vendor experts. Volunteering is really a way to be more committed and connected to the vendor and products. Members are looking for that trusted source to learn from along with the different learning opportunities to participate in like conferences and events. The user group will definitely be providing information and presentations in addition to the vendor products. It comes from real firsthand experience. Critical thinking as part of the leadership skills is vital as well. Whether in IT or a business, role-thinking through problems and trying to analyze the situation would be a key skill to continue to develop. Education about the vendor and the focus of the user group along with the other areas that are important create additional value. The members will benefit from the great education and other offerings of the community.

Voice to Vendor

Interaction among the vendor, community experts, vendor representative, and product managers provides great benefits to the user group. Membership in the community allows for a bigger voice to the vendor. This might even be perceived by the vendor as a benefit to them, because the feedback is important.

In technology groups the members have a list of items they would like to see as enhancements and features. The vendor would definitely be interested in hearing about this. There are opportunities to show how the products are being used and what else is being required to provide what the company needs.

In user group forums details can be provided about new features and enhancements. Example use cases and why a feature is being requested is the supporting details that are needed to give them As members of the user group, these forum groups are available for their participation. The forum can pull together the top requests and consolidate the details that community would like to see. This gives one clear voice for these enhancements and features from the user group to the vendor.

The user group can offer a couple of options for gathering enhancements. A forum can include a couple of selected members that want to be part of the group. A representative volunteer can communicate back to the vendor or the vendor might want to attend the sessions hear the responses. The vendor might even have a way to submit enhancement requests, and that can be encouraged by the leadership to the members. Reminders or pulling together these requests to submit in the vendor provided method could be a task of the user group. The members also get the advantage to know what recommendations might have been gathered and made to see the support of the vendor.

Advisory councils are another way to provide this direct feedback to the vendor. If the vendor participates in the discussion forums that is close to how an advisory councils can work. Members can be recommended by the user group to contribute at the advisory councils. The advisory councils normally start from the vendor and are vendor driven. They might have a specific focus or want to vet a particular feature. Using the members of the user group, the leaders can help if there is a group that needed for a specific area. They can also make members aware of the councils and how to work with the vendor.

Advisory councils could be invited to have a session at the user group conference or event for easier access to members. There are formal issues concerning nondisclosure agreements that will have to be handled by the members but again the user group can either select members or drive awareness for the this opportunity. This type of feedback is so valuable to the vendors and the members become even more involved in the product and feel they have the ear of the vendor to supply them what they need.

Membership surveys are ways to research what they members want to hear about, and they can also provide information to the vendor. If taking a membership survey for the user group, this information can be used for a voice back to the vendor from the community. Outreach to the vendor can even be done first to see if there is something that they really want to hear about in order to include that with the survey. Vendors might also create a survey to poll the community for details that they want to hear about. These might be questions about the product or current projects. Just like with any survey of the membership, it should be limited as not to continually disturbing them and asking for their information. If the vendor is asking for the survey, the best way to have it to come directly from the user group and then the results can also be used for the user group activity planning.

Again for technology user groups, the products coming out from the vendor even going through quality control there are use cases that members have that are different. These different use cases are perfect for beta testing of new products or new releases. The vendor does carefully select those involved in beta testing, and it is not always easy to be included. They do come to the volunteers of the user community and the experts of the community. This is ideal membership feedback. Either the volunteers can gather member use cases or they find some members that might have the edge cases to add the value to the beta testing.

The community might have a call to see which members would be interested in the beta testing. As a member the beta testing allows the preview of the newest and latest features, providing that step ahead of others. Using the beta testing as almost a proof of concept to get ready for the new version can allow for earlier planning of upgrades and migrations. The variety of members provide a more complete beta testing and the members benefit too. Just like the advisory council and learning new products and discussing potential confidential information, there will need to be agreements in place for the vendor so help might be needed to make sure that the member is providing this instead of it being placed on the user group, but with those agreements in place definitely a good experience for members and vendors to learn from each other and improve products.

Beta testing is important to be able to catch all of the edge cases and see what situations that the product might be stretched out of its initial scope. I have seen this in a couple of the beta testing areas, that something was done that was completely unexpected. This provided extremely value details as to what the user did and why they did what they did. Once that was understood, the situation can be corrected either with proper documentation and education or fixes to allow for that or to prevent it from happening. The fixes from the vendor perspective is going to save them a ton of support calls, all from bringing in people that are glad to be part of the community and involved in the testing.

Member to Volunteer

Some of these opportunities we have discussed are with volunteers and that you would need to be a volunteer to participate in the activity. Even though volunteers are members of the community, they have already taken an extra step to be more involved. Volunteers are active members of the community, and they are great candidates for any of the activities with advisory councils and beta testing. They are even going to find that they are going to be happier with the product because they are aware of the direction and with the council and possible beta testing have helped and shaped the products.

Once a member does decide to participate in something like beta testing they are now have been hooked as a volunteer. This might be the one activity and there might be future ways to engage them to keep them more involved. That is going be important for a more engaged community, and that is actually having more people to step and volunteer and have that deeper connection to the user group. The feedback to the vendor has been an easy area to get more involved in because normally everyone can find something they would like to tell the vendor about and let them know what they wished the product could do.

Communication with members is important to include them in the activities or at least to let them know about the current activities and tasks. As they begin to find the areas that are of most interest to them there will be members that will continue on their own to volunteer and start to get more involved. But there also needs to be additional correspondence with the members to promote the involvement and the full benefits of volunteering. They might feel time constraints or not able to contribute, and further details about membership access to the vendor to tie it in with a volunteer opportunity they will start to experience it.

Sometimes it is just difficult to build a community and have members join, and we will be discussing the challenges in the next chapter, but for user to be connected to the community this next step of owning a small piece by volunteering is essential. Membership communication is so important and as the member understands what is happening in the user group it will be easier to include them in these activities. Soon they will be sending out information to others how they participated in the different events, councils and discussion forums.

Summary

The community will not be much without members. In creating a user group, it is essential to have others who want to be involved in the user group. As more members start to participate, the energy and excitement will be experienced in the user group. The activities will have good participation, and communication with the membership will be easy to let them know what is happening with the community.

User group members are looking for reasons to join the community; activities, events, and educational opportunities are the main reasons to join. Education should be user-driven to provide the experiences from others. When starting out this might be difficult to have those volunteers ready to go and start presenting.

Members are looking for more ways to learn and understand the real implementations and uses from other users. Providing education and other ways to learn is going to draw members to the user group. Along with a bigger voice back to the vendor for new features and enhancements, there are ways to be involved in feedback to the vendor. With these opportunities, members of the user group are normally more satisfied customers because they have participated in the product development and have a community to reach out to for extra support.

CHAPTER 6

■ ■ ■

Challenges

So far we have discussed the main areas to setup a user group, recruit volunteers and attract the members to the user group. It is not as easy, just thinking that a user community is needed for a specific product or vendor without doing your homework. A few of these challenges have been mentioned along the way and in developing a strategy for the user group these challenges should be taken under consideration maybe even anticipated to prepare for them.

One of the main challenges in forming a user group is to even know when to establish one. There might be other regional groups already working on building a member base instead of a national or international group, but there might be room for another region. It might be the only group needed based on customer and vendor details but in most of the vendor spaces a company is normally going to be at least national if not international. Most of the time if dealing with a vendor has its challenges then it might be something worth investigating. Without this due diligence in the beginning about the need of the group or what type of community is needed, there will be significant challenges in even getting the user group off the ground. Members will be receiving the information from other places or not at all.

There are definitely challenges in just the normal day to day of a company, and now have a group of users forming a community it will add to the challenges. It is worth working to get over some of these hurdles because of the value and benefit it of the group, and hopefully the fun and relationship it builds too. Let's discuss some of these hurdles as these are going to be stepping blocks to get over, around or through to be able have a successful user group.

Volunteers

There are just challenges in working with an organization that is made up of volunteers. There are several things that work very well at having volunteers but other times it is difficult to get responses or work done because they normally have a day job. It is hard to motivate or hold responsible when they just have volunteered for something. However, reminders of the big picture and advantages to participating in the community have assisted in inspiring them in the best way possible.

Working with volunteers allows for additional enthusiasm and passion because it is something they have chosen to do and help out with. Extra activities that are job related means that you have people that are excited about the focus area of the user group but that excitement might be driven by them wanting to push their own agenda. This is definitely a worry when working with consultants and volunteers but the governance is to help with this if defined properly. Governance will state that as volunteering and part of the group meetings, that it is about the user community and not to drive an individual agenda. If it is part of the culture, board charter or other governance for the user group, it is easier to head off these conversations by pointing back to what was agreed upon and not making it personal. This is a typical example of something that happens and brings the tough conversations to the board room, but also how you can be prepared for these situations if thought about while drafting the culture and governance for the user group.

Volunteer challenges will also have to deal with time and some of the timelines for activities might have to be extended or making sure that the expectations are set properly for the amount of work and volunteer time. Additional communication and follow up is going to be necessary to coordinate with volunteers, members and others involved. Holding volunteers responsible for what they said they would take care of is going to be important, but also having rewards helps. Instead of needing staff or more volunteers to follow up to make sure everything is completed, another suggestion would be to plan a rewards program or little competitions. Rewards can be user group gear, gift cards, event registration just to throw out a couple of ideas. This will not always motivate everyone but it is certainly is a fun way to help give that little extra push that a volunteer might need. It is difficult to juggle all of our tasks and get everything, and volunteering does benefit the user group but also the volunteer. Encouragement and communication will help manage volunteers and hopefully avoid too many issues when dealing with these challenges.

Finding volunteers is another problem. If this is a new group, it makes it even more difficult. There is so much to do and not that many people sure they can know if they want to be initially involved. The membership base is also not as broad as a group that has existed for a while because the new group is still building members, but there are specific opportunities for the volunteers that will help let people know how they can develop the group. I realize that I have mentioned this a few times already, but once the volunteers realize what is being offered, it is not as much of an issue to keep them involved. The problem is knowing where to fit in where the leadership will know the bigger picture, members are not going to see all of the tasks or things to do. The tasks from small to larger for volunteers is going to help in talking to with members to get them involved in these areas. It also helps current volunteers to recruit new volunteers and work on succession planning. If new volunteers are not recruited, people who have been volunteering for a long time can get burnt out or have no opportunity to move into other roles. This doesn't matter if it is a new or older group, it is important to keep offering new people to participate. New faces in front of the user group give other members the vision to be able to be there some day because with the same people it would appear that there are no other opportunities. Finding volunteers just takes reaching out, and developing a volunteer engagement program should be part of the strategy.

Strategy and volunteer planning will help meet some of these initial challenges that will be faced as part of the user group. Other challenges need to be thought through, and probably can be handled differently depending on the group and focus areas, but this will give you examples and help you plan for others.

Priorities

Why is priorities part of this challenge of setting up a user group? The priority should just be to take care of the members, right? Priorities might normally just be set by stakeholders or customers, but there are a few influences, even beside the members, outside of these regular sources that effect priorities. The vendor, members, volunteers, board and others in the environment all have priorities and their list of things that should be done for the user group. Unfortunately, it is not all the same. The vendor might have a new product which they will like to tell the user group about and have them rally around that. The membership might want to just understand how to maintain the current product and how best to use it, and not even be ready to learn about new features. Volunteers might be excited about another event or thing that there are too many and just want to be able to write or talk about another topic all together. The board is looking to grow the membership and planning strategy. If these all aligned or the user group could just satisfy everyone's list we would be living in a dream world. Just as it is with projects and a regular business day, priorities need to be set.

Strategies and plans need to be aligned to help set these priorities. Then the priorities need to be communicated in a way to show the other groups what they are so they are not constantly competing. Part of the communication plan here, especially back to the vendor needs to set expectations. An example of that, a vendor might want different special interest groups setup because they want to be able to talk to a set group of people or they might have a product that they want to have a focus group around. This is great because they are wanting to use the platform of the user group and the user group gets new information as well has vendor involvement. At the same time, setting up the special interest group requires volunteers and staff time. It needs to have other resources, instead of focusing them on the decided priorities for that year. This can cause other projects or programs to be changed or removed for the year. Instead of letting a new special interest group requested from the vendor derail the plans for the year, set very specific guidelines on how to setup a special interest group. This should include something specifically having resources available to setup and manage the group. There should be criteria on how to decide if it should be something that belongs in a new group or should be something belonging in an existing special interest group. There are reasons to have another group, but having to bend of shift priorities because of non user group requests, this is something that needs to be accounted for. Communication about the guidelines is going to help with getting the reason for the request, information about the focus area and providing an opportunity to talk about existing groups. This does setup to have discussions around new special interest groups, but it can be distracting and take away from different activities that are already happening. This is just one example of how a suggestion of a new special interest group can be different than the priorities, and proper strategies can help with this issue up front.

It is a difficult task to set the correct priorities. New products and features roadmaps are going to something that even if not available in a current position a member should be learning about. The user group needs to provide research that is not even yet in the environments. New trends and industry standards are going to help to measure against. Here the distraction is chasing something shiny. It might be the hot topic, but none of the user bases is going to use or it does not quite fit in this user group. Topic might come and go, and there are going to be a few of these shiny objects that are going to make it worth the investment. Volunteers or board members should be listening to the members, vendor, and other volunteers.

Priorities are going to be a combination of a few things; current topics, membership needs and vendor objectives. These areas are going to have influence on the education, events and other activities of the user group, but decisions need to be made with the right combination. Figure 6-1 shows a way to do a brainstorming session on priorities. In the brainstorming session, the leaders can just start throwing out ideas for topics and possible events or other activities. The ideas then can be placed in a chart based on priorities and effort. The effort is the time of volunteers and staff or other effort to make that idea happen. The priorities are ranked from low to high, and voted on by the group so that all agree about the rank and effort. This shows a picture of what is important for the group and how difficult it might be to implement. The example shows a couple of dots representing ideas that are circled, which are high priority and low effort, so these would be something easy to accomplish and include. The ones that are crossed out, are too much effort for a lower priority. Then the debate begins for the other dots/ideas. The resources will need to be evaluated to see how many of the ideas can be included in the plan for the year. Not only do volunteers have limited resources, user groups as a whole do too. In evaluating the available resources, different items might have be removed from planned activities or even strategies. The exercise starts good discussion and assists in looking at the brainstorming ideas from the right thoughts of priorities and resources.

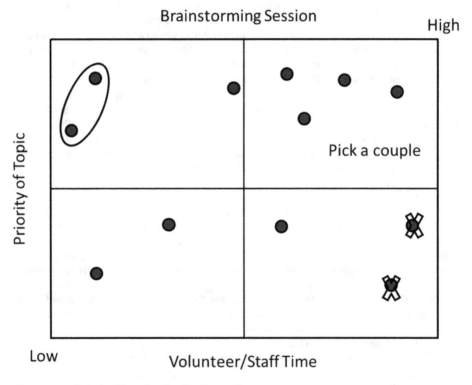

Figure 6-1. *Priority Planning Session Example*

These type of exercises to get to the priorities is a good practice by the user group board. There will definitely be enough discussion generated. Vendors and other volunteers could be included as part of the brainstorming sessions and the board can prioritize and match it up with the strategic plan to make sure it is all fitting together.

Independence

User groups are related with some common thread. Something that group is doing or using, like a product or vendor. If the vendor started the community, it is difficult to break away from the vendor. The vendor relationship is very important in such a configuration. Vendor input needs to be included as part of the community because it is the main thing that the community does. However, it might not be the design of the user group to be independent of the vendor, and they might always be a vendor community. It is for the user groups that are by creation are meant to be independent of the vendor. They would become a separate identity and even with the vendor participating in events and giving presentations, the user group would work providing more user presentations for and user-driven content. So the challenge is to include more of the user case studies, and be able to learn from them as they have struggled or been successful with the products. It is also important to see how everything fits together. So when the user community can network and discuss they normally find something that is extremely valuable for their own environment. The view of the other products and how to integrate might not be provided by the vendor. This is where independence comes in. Independence does not mean to partner up with the competition and give them equal opportunities. The vendor is important and probably sponsors the different events. The vendor that started the user group, they probably receive a prefer attention but based on sponsorship levels.

It is not just the content that designates independence but also the support and where that is coming from. The support is if the group can stand on its own with additional funds coming from other sponsorships and conferences and not fully funded by the vendor.

It seems like it would be easier just to develop the community with the vendor and support their objectives, but in understanding how the vendor products works in the environment actually help the users know how to integrate and use other tools and have a better understanding overall. The independence allows for learning and instructing in these different areas. That is why this independence is a challenge. If it was just one vendor user group completely sponsored from the one vendor it would be dependent. Again there is nothing wrong with a vendor run user group but it would not include as many options and other vendors that might not be partners.

There are challenges with both types of groups. Independent groups might not get the full support of the vendor at events, and will need to explain to other vendors why to support and sponsor a different user community. The support will need to come from several independent sources and there will need to be more focus on use cases and other user presentations, which is a tie in to get more volunteers to be speaking and participating. Dependent groups might have a heavy focus on new products and more marketing centric. There is still opportunities to provide how the products are being used and user case presentations and help with troubleshooting, but the conferences and events will probably all come from the vendor with just including the user community to help with some of the content.

51

Competition

There is nothing wrong with a little competition, but for a user community this can definitely be a challenge because of the limited time we all have. If there are times for other vendor and competing user group conference it will pull members away from one group or the other. Competition means that better marketing and events have to be offered to be able to win over members. With unlimited resources it would even be a challenge, but volunteers and funds are not going to be unlimited.

In understanding the other groups that might be offering the same education or working with community, the user group can either partner with the group or offer different focus areas. Partnering is another challenge because now there is more than one group trying to set direction and planning, but sharing resources so that both groups can benefit would be a way to not only save on user group resources but provide additional content to members of both groups. Before partnering with another group, consider the following guidelines:

- Validate that the partnership will be of interest to both groups

- Make a plan together

- Have an agreement in place

- Divide the tasks so each group uses resources

- Share registration and sponsorships fairly

- Market together

- Start with a small event

Some of these guidelines make perfect sense such as planning and marketing together. The agreement in place will help with defining what fairly means for sharing registration and sponsorships. Starting small is going to allow for a test of the partnership and how the groups work together. It can validate the size of the partnership for sharing the registration and sponsorships. It is possible to go into a longer agreement for multiple events, but a challenge with the partnership might be discovered with a smaller agreement and help avoid something more long term. The vendor will probably also appreciate this when the groups come together to only have to attend one event instead of multiple ones.

Competition is going to be there for just information as to what might already be available on-line and other events. The challenge is providing something unique or only something that the user group can provide. The experts of the community might be one of the options, but they are also offering up blogs and other articles. The opportunity is to either present something first or have the experts save something just for the user community.

Providing an experience for the user community that is going to be different than the competition is going to be a challenge, especially against the vendor community. Trying to be everything to the user so they only have the user group to go to will be even more difficult, but provide the special topic, having a group of experts or pre-releasing information to them will provide a competitive advantage. Partnering might be a possibility to provide more of the scope for users, and using this as an option will provide awareness of both of the groups and additional ways to grow and overcome a few of these challenges.

Events and Conferences

The user group is going to provide a type of event or conference to present the materials and education. Conferences alone are challenges and there are probably full books on successful conferences. This is also where the expertise of the association management company would be extremely helpful, because they have their own event management team and can provide support in several areas. For the leaders of the user group the challenge becomes what type of event and what type of sessions are going to be most useful.

Even with all of the online content, webinars, and virtual type of events, in-person events provide additional benefits with networking and meeting others in the community. There is still a place for these in-person conferences. This should be evaluated for each user group if that applies or if they need a different type of event.

Decisions need to be made about the types of events to be offered. Here are a few suggestions for events:

- Full-week conference

- Two-day presentations

- One- or two-day hands-on labs

- Virtual event offered with in-person event

- Evening presentations

- Vendor and user–combined events

For new user groups a vendor and user combined events might be a good opportunity to develop speakers and engage volunteers. This does provide good vendor context wanted by the community and a story to demonstrate how something is being used.

A full week conference requires quite a bit of planning and is something that will be difficult with just volunteers. Volunteers are required to help plan content, schedules, and the details that will make the conference a great experience for the attending. The association management staff can make sure the entire venue is setup and booked by looking at the size of the attendees and needed meeting rooms. There are logistics with all of these events, which does provide less of a challenge when working the association management company, which of course I recommend. My thought behind that is the volunteers are experts with the products and vendor, they should be able to focus on that. The association management company can focus on the logistics and setup. Other vendors and sponsors for the exhibit hall can be a combined effort. The community has some of the relationships and can build on those, while there are sales teams supporting the effort. It would seem in dividing these areas and allowing the conference to run with the needed support of the volunteers and staff this would not be such a challenge. However, there are concerns around the amount of volunteers needed for speakers and deciding the presentations, and making sure that there are enough attendees and sponsors to allow for an event to be able to support the user group activities.

Matching the activity with the community will be a challenge for a new group as the customers are being understood and how the community is going to react to certain events. Polling might be one way to get the details or just engaging some of the volunteers for their recommendations. Again for new groups the two day events and possibly evening events are going to be less complicated to hold and can have the right mix of user and vendor speakers.

Call for papers and presentations are a great way to get the submissions for conferences. Also keep in mind inviting people to submit gets the right subject matter expert at the conference. Reaching out and recruiting speakers on top of the open call might be needed. The worry is either too many or too few presentations or not enough on an important topic. Matching up the topic areas with the board strategy and priorities might pose a little challenge if presentation submissions are all in a different category. The too few presentations or not the right subject matter will allow for recruiting and it can come from a volunteer or vendor.

Getting submissions for sessions will also help plan the sessions that will be part of the event. If the right education and sessions are planned the attendees will have a great experience. Here are some sessions that you can plan with a two day event or a full conference:

- Technical presentations

- Panel discussions

- Hands on labs

- Deep dive sessions (two to four hours)

- Sponsored sessions to learn about vendors

- Roadmaps (Vendor)

- Use case stories

- Debate panels (present two sides for discussion)

- Keynote (main speaker)

- Flash talks

Technical presentations can be on any of the subject matter for the vendor that is relevant. Sessions can be a how to or experiences as well as what is being used in the environment and why. It is normally not difficult to find speakers from an established user group. The problem comes when it is a new feature or a new user group where it is something that no one is sure that if fits in. But realizing this can allow for two speakers, vendor and user or two different users for support. Panel discussions are more three to six people to answer questions and talk about their experiences in a more relaxed format. Again another way a new speaker can start to be involved or recruited for future presentations.

Hands on labs are not as easy to setup as one might think. There are additional resources that are needed from setup to actual equipment. Some needs to have the experience of lab or actually build it as part of the presentations. This will be time consuming and if the expected resource does not finish in time for the conference what will be the backup plan?

Issues with vendor roadmaps and sponsored sessions are convincing the attendees that these are valuable sessions. Low attendance in a sponsored session can discourage the vendor from being part of future events and roadmap sessions show that users might not be interested in material from the main vendor. If planned correctly with the vendor the content can demonstrate how to solve a problem. Future direction is also valuable to plan architecture and planning for your environments. This might need a little more marketing and right messaging to promote attendance.

Debate panels allow for different sides of the discussion to answer questions and propose another side of the story. These are normally meant to be more entertaining, because even though there are more than one way to do things, there is normally a way the winning team should be. For example, to upgrade or not to upgrade, can be a debate and with presenting both sides, the discussing the issues on the opposing side and how to overcome them. The audience can get involved, ask questions and even vote on a winner of the debate.

Flash talks are just short three to five minute quick tips to talk about a very specific topic or query. This is to involve several people that might not be ready to present a full session and so they can talk about one little problem they solved or how they used a new feature. The quick pace and changes in subject will keep audience engaged and provide examples of some way for them to get involved.

Conference and events should be educational. They should provide attendees opportunities to network and learn. The challenges by events are always the when and the what. When should the conference be held and are there any competing events? Staying away from the vendor conference or event would be recommended but not always possible. Then the what are some of the challenges that were already discussed above with the type of sessions and topics that should be presented. The strategies, priorities and board of director plan should help guide any conference and without that information it will more difficult to engage volunteers for planning and speaking at the conferences. The conference and any in-person event should also have a fun aspect. Enjoying the time out of the office to learn and network with others is normally enough, and too much emphasis on this will distract from the learning. Themes are fun and help focus a conference. People can identify with it and enjoy activities that planned that follow a theme, which would be used for networking opportunities and possible special education events. User group conferences are good for building the community, and they definitely come with challenges that can be addressed with a good strategy and plan.

Engagement

Events, activities, and different aspects of the community are there to engage members and volunteers. The more engaged members are, the more they are going to get out of it. Members that see the benefit of the education and attending events, will be looking for more opportunities to be more involved and possibly looking for a way to volunteer. Volunteers that have started off with smaller tasks might be looking for something else to do.

When members and volunteers are engaged the user group and community is going to grow and the feedback to the vendor is going to be more relevant. Involvement and interaction is going to be providing valuable insight and help with product implementation.

So, why is this in the challenge section? For some, being engaged in the community is going to be easy because they might have been involved before in a user group and seen the benefits. For others they are going to see something else they have to fit in their day when they really just wanted to know how to do one little thing or join for the conference. Volunteers might already be over committed because they have said yes one too many times and getting information out there might be difficult to keep it current and relevant.

So for members and volunteers who are too busy, how do you keep them engaged? Now you are starting to see the challenge. The leaders of the user group can start by communication plans. If there are regular plans that go out about events, activities and everything happening in the user group members might come across a few things of interest.

Gamification might be a way to draw in members and if you are using social media tools, help contribute to discussions. It would be important to include in the strategy keeping the education relevant to things that are happening and changing with their environments. As you can see there needs to be strategies around the tasks and activities that are available, then communicate that to the members. Those that jump in a start to participate should experience something that is awesome and allows for them to bring another member along. The user group leaders have a responsibility to reach out to members and volunteers to encourage them to step up and enjoy being a part of the group.

Sometimes this will be the easiest challenge and sometimes it will be the most difficult because when people are not getting involved the strategic plan and priorities will start getting questioned. Reviewing the strategic plan to keep focus and ensure the engagement of the community will verify that other programs following the priorities and plan are still aligned. Engaged community does have better discussions and input to the vendor and others in the community. A select few will not be able to carry the whole group, so there should be thought put behind activities and involvement of members and volunteers.

Summary

There are a number of challenges involved in starting up or leading a user group. Priorities need to align with the strategic direction of the board of directors. However, priorities can be derailed by those involved in the community such as volunteers and vendors. Feedback is needed from these different groups to make sure that the topics and education are in the right priority but the user group leaders should be validating the priorities and providing communication instead of others changing priorities.

Events and conferences take a lot of planning and come with their own set of challenges, but they provide in-person events for the community to meet, network and learn. Engaging the volunteers will mean enough speakers for the events and members that want to attend. Vendors will also be able to participate in an independent user group event and provide sponsorship. The vendors have opportunities to present how their products complement and fit well in the communities' environments.

Independent user groups have a strong voice back to the vendor, but the purpose of being independent is not to oppose the vendor but work together. It is also not to just include competition in the community but to understand how all of the pieces work together. Being independent does allow for this and comes with the challenge of balancing the independence and getting the needed support from the main vendor.

■ ■ ■

CHAPTER 7

Techie to Leader

I have been in a technical career for quite some time. Problem solving, querying data, writing code to automate something, and testing to validate that it works makes it fun to be in technology. There is so much to the database world with tools, data integrations, administration, development, and, of course, security. Database teams work with several partners and there are interactions with different teams. There are technical issues and problems to solve, and while that might be good enough there are people who would like to be leaders, and I am not talking about just leaders in their field. There are leaders in their field and there are leaders that also have the titles to go along with it. Finally, there are those that can continue with all of the technology fun and demonstrate leader skills and especially considered a go to person because of that. There are ways to develop these skills that get you noticed in the organization, as someone with the technical skills and abilities to make the important decisions.

There are plenty of other technology jobs that have the same level of the technical skills. Technology career paths continue to be complicated, and even if said, you can stay technical there is still a level that must change or remain a techie. Hopefully, you have figured out by now that is not the case. That even in these different roles and careers that there are opportunities to develop leadership skills. These are definite soft skills that you can develop, but how do you actually get out of technology to move into a more defined leadership role? I believe it is already demonstrating leadership skills in the technology role. Leaders of a team lead with technology and by enabling their team. The following chapters will take a look at the qualities of a leader, but let us first look at the potential that we all have to become an example of a great leader.

Examples of Leaders

There are plenty of articles, books, and blogs on leadership available. There are examples of leaders from CEOs of many different company, influential people, and even from Captain Kirk. As you look at these examples, a list starts to form about what we understand to be good leadership qualities. Notice that these are not just managers or CIOs but they are people that have the leadership qualities and challenge us to do awesome in our tasks and jobs. An example is definitely needed to be able to learn good leadership development and even the bad to avoid certain behaviors. The difficult part is finding a good example. Also, the people that we tend to look for are those that are put in charge of people, companies, spaceships or another group. In a user group we already have people providing strategic direction, leading and providing sound technical guidance.

The leadership of the user group can be planning and giving leadership guidance to the community. This is a perfect case, because the technology experts already know and understand the technology and some of them lead their company, teams or even user group. Even some of the leaders of the user group might not have that role in their company, but have been demonstrating leader qualities on their team and with the technology that they have been using. We are going to find people that have qualities that we are expecting from leaders to use as examples. The user group can actually provide mentors and mentor programs for additional offerings for leaders, but it can provide the strategic leadership training.

An example of a leader coming from the user group is a previous manager of mine. She was and still is involved in the user group in several different roles. She challenged me every day at work to think past the current problem or issue to what other questions or problems might arise. She gave me opportunities to get involved in the user group and find a place where I could also fit in. I was able to grow into leadership roles in the user group because of this first push and encouragement to be involved in more than just a community member. This taught me to continue to grow and work on communication skills.

Even if you are looking for how to be a leader from Yoda's example, understanding the characteristics that attract you to that leader are going to be the areas for development. Some of these skills are developed naturally from being in technology. There are advantages from the day to day to build these skills and help move you toward a leadership role. User groups can also support these growth areas with sessions on skill development and how to move and grow into a leader from the technology space.

Technology Teaches Leadership Skills

In technology, there are skills we learn that are considered key leadership skills. Any of the groups in technology can work in silos and just provide a service and not think about what others are providing. For example, the network team can set up network systems without taking into consideration what systems and data are going to be passed over the wire. If the network team works with the server, application, and database teams to understand the traffic and get information from the stakeholders, the service is going to be what is needed and more productive for everyone involved.

More Than One Way To Do Something

With database systems there always seems to be more than one way to do something. If asking a question about how to do a task, it is typical to hear an answer that starts with "it depends." That means that there are options. Planning and reviewing with others to get the best solution for the environment in place allows for collaboration and accepting different viewpoints.

Diversity brings different questions. When trying to solve a problem or plan for a future project, asking different questions leads to different answers and overall solutions. Leaders do not necessarily have all of the answers because they can provide ways to help the group get to an answer by asking questions or engaging the team to bring their expertise to the table. The solutions and environments that come out of a team like this would be a collaborative effort with the best direction and solid solution.

User group communities bring diversity automatically. Accepting different opinions and thoughts in order to organize a solution is a leadership skill. Allowing different ideas to come together is important in a team lead role or in on a board of directors for the user group. The board is going to have several different opinions and thoughts and to lead the group you need to have skills to directly communicate, understand and respect the different options to be able to come up with the best direction for the known issues and plan at the time.

Constantly Learning

In technology we are forced to learn. There are upgrades for existing platforms and new technology that gets released or changed. Applications are change to handle new enhancements. All of these and more make us learn so that we can continue to support the environments as needed. There are plans for future growth that also happens in support of these systems and in researching roadmaps there is plenty to learn to see if growth and new features are going to change based on these roadmaps. There is a constant learning that is absolutely needed for any technology professional.

Technology and company environments are on not going to stay the same. Skills are to be developed to look at the new areas coming down the path and find a direction that will be beneficial to the company and it is also something important for the user group community. Changes are going to need to happen and sometimes change is more difficult to lead than looking for a future path and course correcting along the way. Managing change is a leadership skill that is going to get a team behind a change and be supportive and in turn championing the change.

With user groups change is necessary to keep the group relevant. There might need to be a new strategy, different way of presenting education or changes to conferences and events. The board and volunteer leaders are going to need to manage these changes. Understand that there will be resistance to new ideas and changes.

Learning leads to growth and changes. Technology professionals will need to continue to learn to manage the changes in platforms, upgrades and new technology. Leadership skills are developed out of learning and managing these areas that come from moving forward.

Need To Ask Questions

Seeking to understand is definitely one of those rules that I live by when going into a new environment. This is something I believe any good leader would do as well. Running through an environment and changing everything to the way you want or believe that it should be will not allow the people that were already there to be too eager to follow. It is a skill to come in and ask questions to understand while still having the appearance of understanding in the first place. Intelligent questions will help to see why a system is configured the way it is and still thinking through a process to improve it.

Leaders are going to be able to ask the questions that are needed for requirements as well. Gathering requirements again for a technology professional is something that is going to happen on a regular basis. Stakeholders need to provide their requirements and

I am pretty sure I have not understood all of the requirements on the first time around. There are always questions that arise. Going back to get the questions answered based on the technology solution with the needed requirements.

The other area to ask questions goes back to more than one way of doing things. To be able to have discussions to make everyone able to contribute and feel that their opinions matters, asking questions first to understand is a good way to handle this.

Anyone in IT has learned to ask questions. They might not always be the right questions, but the questions are needed for the requirements and to understand what the stakeholders need. Questions are the best way to grasp why things are the way they and to really how to work well with the other teams. As a database professional the questions are going to be around the right data and how to analyze the data, and these questions can transfer over to business analysis skills. Asking questions of other teams understanding how they work together and building the relationships are skills that are needed for leadership roles.

User groups are also a great place to ask questions to gather other opinions, find out about different options and discuss strategies. Using these skills in our technology field and in the board room or as volunteers strengthens the skillset and develops it in order to be used in manager and leadership roles.

Communication

Asking questions is not the only way to communicate. Technology teams need to communicate operational tasks, production issues, and project statuses on a regular basis. In response to troubleshooting tickets there are needs to communicate back to the owner when they are going to be able to look at the issue and set expectations of when it might be fixed or reviewed. This type of communication is critical between teams and applications owners. Communication is an extremely vital skill even if not in leadership, however, it seems to be one skill that we can all do better.

As a technologist we develop frameworks to make it easier to communicate in a more consistent and standard way. There is definite value in these technics for standardized communication because it becomes easier to do and in a format for everyone to understand. In receiving these regular communications the visibility is increased and the bar level for communication has been raised.

As a leader able to communicate to a team to establish goals, direction and enable team members. Communication to members and volunteers needs to be in a consistent, transparent and timelt fashion.

Making Decisions

Decisions in technology come in a few different forms. There are technology recommendations, troubleshooting implementations, and production issues, to name just a few. Evaluating new technologies and products are a part of the job. Due diligence on the various products to validate if they meet the use cases and make recommendations for the proper selection is also vital. Decisions are made to implement the best possible tool, and leading a team requires making decisions team and management related.

Decisions are being made with production issues and troubleshooting. Working with systems and understanding what went wrong allows for quick thinking and presenting the best course of action, so making the decisions to remediate the issue as quickly as possible. Quick decisions and actions that are decisive are developed by being in a technology field.

Decisions are also made from implementations. If designing and planning is being worked out there will be decisions that will need to be made to be used to implement it.

Making decisions is a valuable skill. Managers will then lead the direction for decisions made by upper management or from the board down to volunteers and the wider user community. Users and volunteers should feel enabled to provide recommendations for good decisions.

Why Not Both?

Developing leadership skills from technology roles leads to different opportunities. Normally if leadership skills are desired it might be expected to have a management role to grow into. There are also other leadership opportunities that might be available depending on the company. It might not be a desire to move into managing people, because the technology is the fun part of the job. The thing that you like about the position is the technology and getting involved in new projects and systems. But even in order to get the new projects and the new technology is going to be important to be skilled in leadership to demonstrate communication and persuade others to see why it is valuable.

Just to go into management because that is the next progression of the career or job path doesn't mean you need to give up technology. Learning a new technology or developing a plan to use a new feature would be driving others to use the new plan or other opportunities to enable others to learn the feature and inspire them to be excited about the new implementation or feature.

Skills needed to continue developing technology would include:

- Troubleshooting and weighing options
- Asking questions
- Constantly learning
- Making decisions
- Communication

Staying technical might be the single reason for not moving into management, and that is where it might be worth considering how to do both, in that developing leadership skills and staying technical.

Being able to focus on just the technology and not sharpen any other skills will let even some of the technology side suffer. The considered experts in the technology areas are going to be people that can communicate with others in a way to make them the go to people for questions.

To answer the question, why not both technology and leadership, it really needs to be both in the sense that developing leadership skills is going to strengthen skills that are part of the technical side. Also, it will allow technical experts to be able to continue to focus on the technology but inspire others and become an effective technology leader. Leadership doesn't mean managing or having a manager or director type role but it does mean to demonstrate these qualities and continue to grow. One can do both, absolutely, technical role with leadership abilities.

Since both technology and leadership skills are valuable for the community the user group should be able to support these efforts and provide education to build these full set of skills.

Sharing Experiences

Becoming a leader, especially a technical leader, there are plenty of stories along the way including decisions as to why to stay on the technical side and how to use the leadership skills. The user group is a platform to share and a must have discussion because there are several people that go through the same thoughts on what comes next and how to stay technical. Others will develop the leadership skills and take their technical knowledge with them to manage teams and other technology teams. Understanding how to get both places is extremely useful and the user group can assist in conducting these types of sessions and as we talk about in the next chapter career development.

The community is about sharing experiences, how things were done and what are the lessons learned. Being able to learn from someone else's mistakes prevents from making the same mistake, at least you hope so. Taking the good experiences and again learning from them and how to apply them in order to incorporate the best way to do something.

The user group has more to provide than just the product information and how to succeed with a vendor. The community has more experiences to share and a plan can be created to incorporate these with other education. It will allow for a more successful community as they learn what they need to continue to development with leadership skills and how they can use the community to enable others to learn.

Experiences are another way to get people involved in the user community. They are the only ones who understand their own story and experience and they can offer this as a way to get started with speaking. It is only a story that they can share because it comes from their perspective and it can be on how they handled a difficult situation, developed a communication plan or even how to ask good questions. It does not have to be a complete technical session, but can be an experience with one piece or how they are development some of the needed soft skills that are required for technologist.

Developing Soft Skills

Volunteering for the user group can build and develop soft skills. The experiences that are coming from the different users are people to network with, and hear these stories. Speaking in front of people is not a necessary soft skill but it can help develop other skills such as communication and continued learning.

Becoming a techie leader or move over to the management track the soft skills such as communication and working with others are needed. In order to communicate better, develop a standard plan or framework. Even a standardized template to send out to a group of people will help to ensure that the right points are communicated. Having a distribution plan will help to protect against missing someone out or missing crucial deadlines. If we take a look at this example from a user group perspective, the user group has opportunities to communicate to the membership and perspective members. A communication plan will help to ensure that community strategies and activities are properly announced. Templates for members and perspective members along with the distribution list and plan will allow for a more informed community that is engaged. The volunteer that is working on the communication will be developing these soft skills not only in assisting the community but to be used in the day-to-day job.

Summary

Leaders are everywhere in an organization. They do not always have the title that designates them as a leader, but with the skills they demonstrate and how people respond to them. Additional assistance to continue to learn and build on these skills is critical to be successful not only as a technologist but to gain the recognition and expertise level to lead others.

There are plenty of leader examples out there, but carefully choose the right ones to understand both good and bad qualities. The user group has an opportunity to bring leaders to the front as good examples. Volunteers in the user group are building these skills and education can be around the technologies and leadership skills. Leveraging the user group the members are already engaged, wanting to learn and volunteer. Volunteers can demonstrate leadership in several different areas of the user group and then bring it back to their organizations. They will be more effective as a technology leader because of this experience and examples that they have learned from other user group leaders.

If the goal is to advance into a leadership position, working with a user group is a great place to be learn and practice these skills. Technologists can continue to expand their skills that they are doing with the regular day and not all of these skills are a big stretch from a good technologist to a leader. Technologists can stay technical and lead in the technology, education and enabling the team to reach the goals. Or they can educate and pass the technology to other team members to engage them for the projects and lead them in that way.

Technologist provides a unique way to become a leader. With these options they can grow in their careers and develop a plan to advance. This group of leaders can also help bridge a gap between the business and the technology because they have developed the skills to communicate, plan, set strategy and ask questions. In the companies and organizations that are participating in the community, there are better partners and leaders to have better technology solutions to support overall business and enterprise. With the involvement and growth there are going to be new career opportunities and a chance to develop a career path based on leadership and technology expertise.

CHAPTER 8

■ ■ ■

Career Development

I have had jobs and careers. Jobs are a place to earn money, but there is more to careers. Careers are good things and still have an interesting aspect almost just funny. If you think about we work to pay bills and spend time doing thing we like in our free time. I look at careers as something you actually enjoy at the same time, so work and spending time doing something we like is the same thing. It is entertaining to me because I have had a few careers in my lifetime, and have felt the most at home in technology. I have enjoyed some of the other areas I have worked in but I started to see the opportunities in a technology field and switched careers from financial to database professional. I am also finding my career has changed as my experience has increased which tends to make sense. Different experiences have led me to explore these opportunities and evaluate what I am doing. It is a dream to be able to have a job that you enjoy doing, but if you think about it, it is where you spend a ton of time. I am not trying to be a career counselor here or bore you with my career path, but I believe that there is something to be learned here about finding an area of interest by exploring, researching, and exploiting the available resources. If it is something you enjoying doing day-to-day, it becomes the hobby, too, with involvement in the user groups. I have other interests outside of databases for my some of my free time, but because I enjoy my field and my experience with the user group they play off each other to make what I do more interesting and allow me to excel in these areas.

Career development is something that continues throughout all of the positions you have. Development comes from making a plan, establishing goals, and engaging others to help hold accountable to achieve those goals. User group activities should be included in those goals being able to either use people in the community or the tasks that they are looking for volunteers.

If there are tasks or activities that are not currently available in your current position, this is where volunteering comes in. In volunteering you will have access to activities outside which will definitely help with goals and gaining the needed experience. This could almost be considered leadership training or career development training. Participation on the board of directors offers access to budgeting, strategic planning, and managing people. Working on a committee will have collaborative work, project planning, and some marketing. Volunteering to write blogs and articles will improve writing and communication skills. These are just examples of additional skills and experience that you will gain from volunteering with the user group and any of these skills can be used to meet goals for a career development plan.

Figure 8-1 illustrates different things that influence our careers.

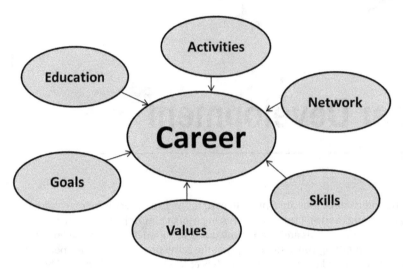

Figure 8-1. *Career Influences*

We do have our goals and values that lead us to join user groups, get involved in different activities, and continue to learn. In the next section we will continue to discuss how important development of skills are when we become a life learning to build out our skills and seek after education to help in further career development.

Learning

Growing and learning are part of development. This is one of the main reasons that I enjoy technology is that it does force continuous learning. All of the new releases, new features, and new tools and hardware help with this. Honestly, just with technology you do not even have to look for a new career, but either get better at the current position and grow into a new one. Learning is such a big part of career development. The user group can provide several areas of education.

As a user group, what can be taught? Perhaps before discussing that, we should look at how people learn

How Do You Learn?

Some people can dive into documentation and learn everything they want to know by reading a manual or book. Others can learn by listening to presentations, asking questions, and absorbing the information from an expert in the field. Still others need to experience it, as they have to see how it works and try it out before they can grasp the concept around the technology or product. Learning can be accomplished by listening to all possible materials, webinars, presentations, and recordings, and researching information to read white papers and other article supplements.

It is typical that learning would be a combination of methods. The initial step might be the research and then listening. After grasping that material, actually doing will allow for the materials to be tested and see if there are other areas that need to understand in more detail. If more details are needed, a network can be tapped into to discuss such as a discussion forum or a meetup to get be able to get a better understanding to continue to do.

But the way that some people know for sure that they have learned a topic and really comprehended it, is to present on it. I actually learn more when I give a presentation, especially when there are great questions that I might not have thought of before. That gives me the motivation to go back and get the answer by researching or reaching out to others that might have already experienced that situation.

User Group Education

All of these ways that were just mentioned on how to learn can be offered through the user group community. The different offerings to have a journal or a collection of white papers will allow for the research and read of the information. Webinars, events, and seminars to present on topics to allow others to get the education on the technologies or products can be done by volunteers.

Figure 8-2 shows how the ways to learn can be handled through the user group activities. The education that can come through the user group volunteers can meet the needs of the community and have a surprising effect on those involved in that more information is learned and shared.

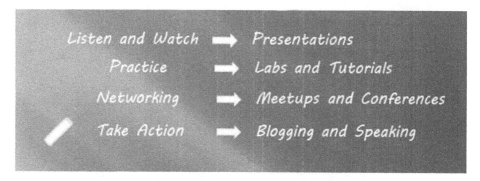

Figure 8-2. Ways to Learn through Activities

Several of these offerings and education options through the user group have been discussed as part of the strategies and activities. However, in discussing career development, it goes back into not only learning but looking for improving various skills.

Skill Development

Lifetime learning and development of skills is going to produce options for careers and jobs can bog us down with the regular day-to-day to not give us the opportunity to spend time different skills. Quick information or having a source such as a user group to get a dose of something new can boost the skills. From the perspective of planning from the user group, having quick tips, regular information blasts, blogs or short articles can provide a wealth of information for skill sharpening. This would be a great value add for the members in the community. Being able to take a weekly email blast or a blog once a week to review to help in this way will provide a way to continue to update the community and help in skill development. This is something I do spend some time doing every week, just researching, finding new materials and the user group news and websites are the first places I go to check it out. Spending this time to see what might be new and what others are thinking about helps to get some additional information and learn something.

Reading and researching a topic is not enough to develop a skill; some experience would definitely be needed. User groups offer this opportunity to learn about a skill and can provide volunteer opportunities to gain that experience. This experience is very important for career development because not only with the skill development you are learning new skills, but also realizing areas of interest. Skill development for weaker areas will help reach career goals since those are probably needed skills.

Figure 8-3 shows possible paths based on management versus technology roles. Starting roles or levels can be the same. Skills can be developed to move up the path and possibly across paths.

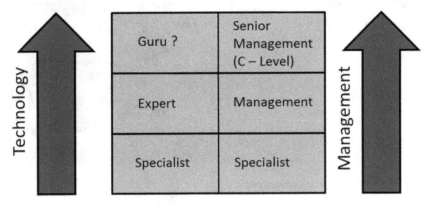

Figure 8-3. *Career Paths*

With leadership skills it should be possible to move from Expert to Management. Even with more technology skills it is possible to move from management to the technology path.

Networking

It might be who you know to get into a career path that you want to be in, but it can also be who you get to know. The community network has experts in the products and experts at their places of work. If they are consultants they have additional contacts. The vendor is also part of this network, which helps to provide additional expertise but other job opportunities.

That was actually part of some of the problem with succession planning is when the vendor decides to step in and recruit people away from the user group to go and work for them. When presenting and volunteering these members start to be recognized in the community and by the vendor. The user can work with the vendor on other tasks and projects for the user group, so there is additional exposure to the vendor. Eventually, the vendor realizes the talent that they have in the user community. The user group did provide the networking and areas to volunteer, and in taking advantage of this the member did benefit and find another career path. I guess when you are doing the right things for your membership in providing ways to continue to learn, skill development, and career building opportunities it would be better just to promote this as benefits of the membership. Redefining the succession planning can be the least of the concerns.

Networking to have contacts and experts just an email or next meetup away has already been discussed as a big part of the user group. Networking contacts can help talk about their experiences and how they got to their position. They can provide the skills that were the most valuable to get there. There will be others that can help out with the decisions of staying technical or moving more toward a leadership and management role. Along with leadership topics, there might be a track to have people share about developing into the roles that they have and what they like the most about the role. Career building means to also understand how to get there and where it is you want to go. Learning about other's experiences helps to formulate these goals and direction.

Even learning about being a consultant versus a full-time employee can come from the community. I gathered a bunch of information here when I was playing with the idea of consulting. I was able to find out several details and things to take under consideration when being a consultant. It is nice to have choices and making friends in the community makes it easier.

Career building in the community might mean the user group can have a job board and resources for skills and other feedback. It is something that other support vendors can also be a part of and a great place if another company in the community is looking for people. The community is a perfect place to look good people because they are already excited about the product or technology, wanting to learn, and building skills by volunteering.

Advantage User Group

The user group played a very big part to my career. When I first started working in the database world, I joined the user community. It started off as a great place to get a variety of education around the technology and database. I started to meet people and they all saw the enthusiasm I had for learning and the databases. This is also where I say networking is about who you get to know. There were volunteers in the community that encouraged me to also volunteer. I picked some topics that were important to my current position and projects we had going on, and I was introduced to different product managers and developers for the products. It was very valuable to get this information for implementation and getting everything setup and questions answered. The projects were successful and I continued to learn and build my career. The next set of skills was leadership. I was able to be in the special interest groups and manage a group of volunteers and lead the group. This provided me a place to practice leadership skills, make new friends. The technology I was learning was being applied in the work place so the company was benefiting and I was learning as well as supplementing my career through volunteering with the user group.

There were opportunities to speak and share what I had been learning. In talking with people I had met in the community, I was offered to help write a book. Things that I could barely imagine doing, but they were coming from getting to know people in the user group community. The vendor also took notice and I was able to start speaking in different places all over the world, talking about the topics I enjoyed and also very relevant to others in the database field. I was able to meet more product managers on the vendor side, and built up even more understanding of the database software. Not only was I building a network of friends, but I was recruiting, too. This was great to include others and mentor them to be involved and learn even more.

In dealing more with the database and the areas of security I was able to add value to my company while still being able to focus on other parts of the database. In having specific database security projects I was able to bring feedback to the vendor on performance and areas of concern. This did help significantly with building these relationships.

I had the privilege of being on the board of directors for the Independent Oracle User Group, which allowed me to lead a group of volunteers and association management staff. As I learned more from the user group, I looked for people in the field and members that might not yet have been engaged, and turned around and did the same thing that others did for me. I encouraged them to speak at the conferences, write for the journal and start to participate in the conference committee or special interest groups. In setting up succession planning with the user group board of directors, I was developing skills that were needed for regular day to day in planning cross-training, having others involved and excited about their role and goals of the team. I was enabling others to do the same. It is an amazing feeling when you are involved in a community that is really working to help each other out, provide services to improve the areas and careers that we are in and develop leaders in technology and in their fields.

When I do look back, I was provided with some great opportunities to get involved and I am glad I was able to roll up my sleeves and learn and volunteer in various ways. Any time I am looking into a new vendor, I look into user groups that are available because that is going to be the best place to plug in and really get to know about the vendor.

Summary

The user group is not necessarily about career development, but when you get involved in a user group you have opportunities to be engaged and develop skills that are not possible from just the regular job. The goal of the user group is also around education and networking and another benefit that comes out of it is providing the community and members chances to build out a career path that might not have been anticipated before.

Since the user group provides the education and additional skills such as leadership skills are valuable and should be included the additional skills can help in planning careers. The user group should take this into consideration and realize that they are also in career development too. The user group can give opportunities for volunteering and a place to gain experience with these skills that are being offered in education.

The user group is not a career counselor, but people that are involved in the user group are happier in their jobs and with the products that they are using. So there are definitely opportunities to support the membership with networking or additional skill building to have better careers or move up through their career path. With decisions that you have along the way in regards to management roles or technology, networking can help to formulate and consider the options for the choices.

There are several involved in the community that are that way because they have experienced the benefits and how it has helped them get access to product managers and other vendor experts. It is a great way to build out a career, plan by talking with others instead of just sitting back relying on things to fall into your lap. It is work to develop skills and gain some experience to show what skills have been learned. Leadership skills are needed for any path taken, even if continuing down a technology path to be a technology group leader it is going to require these skills.

Consideration should be given to make sure these leadership skills and other soft skills can be included in the user group conferences and events as part of the education strategies. The volunteer roles will give experience. Active engagement in the community will allow for you to learn more and want to continue to pass along to others in the community. There are definitely ways to take advantage of the programs of the user group to provide better career development. And the user group should be pursuing ways to have these strategies included.

A strong community will start to promote and be more engaged because of these opportunities that they are given. Members are going to be more enthusiastic about the skill development and learning including the products that they are working with. All of this helps with being more excited about their careers in this field.

CHAPTER 9

Qualities of a User Group Leader

A user group provides a variety of benefits for the members and the leaders of the community. The leadership of the group sets the strategy and develops the plan for growth, and the list of things that only a board of directors can do requires strong leaders. Starting a user group takes a ton of work, and participating on the user group board of director takes time, too. Going into it for the wrong reasons—self-promotion, fame, and fortune—will just leave you frustrated and not allow for the development of volunteers. In order to start to build others up as leaders, you need to understand yourself as a leader and seek out qualities that are going to have other good leaders part of the team.

The number of books on leadership speaks to the fact that there is not one solution to become a leader, and there are several ways to get there. There is not one specific quality that makes a leader but a combination and a willingness to learn. A few qualities stand out as being more important to have. They come from character and values that a person has, and other qualities can be developed.

Leadership might be easier for some people more than others, but people are not born to be leaders. Being a leader is a given opportunity to enable and inspire others. Different people are able to communicate with ease or plan strategies because of understanding the big picture without too much effort. That does not mean that someone can grow into the role, especially with the desire to do so. Recognizing that, there are ways to work on some of the weaker skills and characteristics.

To be able to talk business and technology is definitely a valued asset to companies and with the leadership skills development and looking at these qualities of leader is going to be the way to get there.

Understanding the qualities as the goal is secondly important to understanding yourself and where you are with this development. Reflecting and knowing yourself is going to allow you to discover the areas that come easy and start to work hard on the other skills and characteristics that require the effort. Maria Anderson, Independent Oracle User Group President, stated in a recent article that, "self-reflection is a skill that, when well developed and practiced regularly, can bring you clarity and help modify your behavior in certain circumstances; in other words, it enhances your self-awareness." Self-awareness will help to evaluate where you are with the qualities and recognize how you are doing as a leader.

Before we discuss the qualities of a leader, let's start with looking at what a user group leader does. This will also help with the self-awareness to realize the areas that are needed and what else might need some development.

What a User Group Leader Does

A user group leader has responsibilities to the volunteers and the members of the community. The leaders work toward the common goals and work to plan for the outlined strategies. Of course they lead, right? The community offers so many opportunities with expertise, mentoring, and other ways to lead. But those put in leadership roles have volunteered to work for a community that they are excited about and wanting it to be successful.

There are wrong reasons for being in this role. Wanting your company and self to be successful makes sense, but it should not be the only reason for wanting to be in leadership of the user group. Fame, recognition and money, not really the reasons for leading a user group either. Self-promotion will cause disruption in the community and make it difficult to recognize if the direction of the user group is planned for the whole community or just their individual benefit. Not to discourage anyone but there is not normally much money to be earned in spending hours in planning and not getting paid for it. These are definitely not the main reasons, because getting to be up front is the easy part, well, at least sometimes. The work has to be completed first, and the overall vision of the group needs to be considered.

So what are the qualities that a user group leader has?

- **Experience.** One main thing is that the leader is also using the product or has that experience. In understanding the technology or product from the day-to-day job, this provides the needed details about what users are facing. Understanding the new and relevant topics for the space helps decide direction and strategies.

- **The Ability Know to Find Answers.** They also realize that because of these complex environments, it is not possible to know everything about the systems. This would help in deciding other people and leaders to surround yourself with because if you can't have the answers it is important to know how to get the answers. They know that the user community is going to have these different experiences and speakers can provide more the complete picture and a well-rounded education.

- **Willingness to Participate in Discussions.** The board and committees will have discussions. The volunteer leader brings their ideas and shows up ready to participate in the discussions. Just listening in doesn't help for the smaller group that will be planning and making decisions. Questions need to be asked and as the whole board discusses nothing should be left on the table. Different ideas are needed or support of current ones which should all be part of the discussions. Other discussions will be where the money should be spent, as there are financial and budget discussions. Without understanding the events and activities with what is needed by the user community, budget discussions are going to be difficult. Reviewing the budget and discussing the financials are a part of what the leader does.

- **Consistency in Attending Meetings.** Attending board calls and conference calls ready to discuss and actively participate. Being able to understand topics for discussion and understand provides more valuable input. There might be times where you have more input than others which makes sense, but there is preparation that needs to happen before the call. Just showing up can allow for the whole group to discuss, but it is important to be prepared to discussion. The number of meetings will depend on the role but the board of directors. The meetings include the board meetings, conferences, committee meetings, and special interest group meetings and depending on the groups you are involved in would be the number of meetings that are on the schedule. Email is good for a few things, but direct conversations are also needed. Also there will be face to face meetings. Boards and user group leaders are not always in the same location, so face to face meetings might not be as many but they are even more important to attend to relate with others.

- **Recognition of Importance of Networking.** Networking with other leaders is needed to create the group and have the plan and strategy. The networking does not stop there, because there are vendors and other user groups. In person events are critical to be talking with others. Networking with the members adds strength to the offerings and building the community. These are the future leaders. Vendors need to understand the direction of the group of what is happening and planned for the group so that they can help support. Meeting with other user groups is key in the networking circles. This helps to know what works, gain other ideas, and possibly plan events together.

- **Good at Cultivating Relationships.** The board of directors has to work together and building the relationships between the leaders helps to understand how best to complement each other. The diversity of the leaders is absolutely needed to pull in the overall picture of the environment. This does tie into the networking, because once meeting the groups of people, the relationships need to be built and maintained. Relationships need to be cultivated with the vendors, other user groups, and the community members.

- **Desire to Continue to Learn.** Constantly learning is the overall theme besides leadership of this book. But a user group leader is out there reading, researching and learning. This is extremely helpful for the day-to-day job, but is needed to verify the direction of the user group and see if there is anything new and relevant for the community to be learning as well.

- **Ability to Make Decisions.** Decisions need to be made about events, speakers, webinars, and so many other details under this. It is the responsibility of the user group leader to evaluate the details, understand the costs, and verify against the board strategy. Based on this information decisions need to be made. Budgets get set and voted on to make sure the plans and spend follow the direction and meet the needs of the community. Decisions get made to run the board and the user community.

- **Ability to Direct Staff.** The staff might be volunteers, but an association management company would be providing staff to take care of the details of the projects for the user group. Regular calls, planning, and managing of the activities comes from the user group leaders.

- **Ability to Enable Others.** Continue learning and enabling others is definitely a reoccurring theme, and as a leader we should be able to bring others along and learn from them. Removing road blocks and providing opportunities to do the same. Letting others look into options, propose solutions, and make decisions while supporting them.

This is not an all-encompassing list of the things that user group leaders do, but it does list a few for starting. Remember that the user group leader is not alone, because there is a whole board for support and hopefully a membership where a few good volunteers can also be recruited to help out. Even if the current leaders might not be available, previous leaders are a great resource and can provide encouragement. I think this is one of the reasons that I am still participating in the user group even without still being on the board of directors. It is also why this information is being provided to you as a resource and hopefully a source of encouragement.

Leaders of user group are involved in several areas and perform tasks outside of strategies and planning. Being able to perform these levels of tasks does take practice, years of involvement, and a desire to improve skills and develop as a leader.

Qualities Outlined

As already discussed the skills of leadership can be learned and developed. There are qualities of a leader that show the character of the leader. The best way to find out about qualities of a user group leader is to ask current and previous user group leaders. What was interesting in asking the leaders what was one quality that they felt was most important, they all could not just give one. If they had to choose they eventually gave one, but there were specific reasons and still felt that multiple qualities are important. That does mean if there is a weak area that is being developed, some of the other qualities will help with the progress and further leadership skills.

I sent a quick email out to a group of people that have been very important in my development as a user group leader and asked that question, "What is one quality you look for in a leader?" Polled were Maria Anderson, current president of the Independent Oracle User Group (IOUG); John Matelski, Ian Abramson and Rich Niemiec, past presidents of the IOUG; Blake Wofford, current president of FUEL Palo Alto Networks User Group; and Carol McGury, previous interim executive director of the IOUG and executive vice president at SmithBucklin.

Maria said, "I'm not sure I can choose one quality but I can say my top two (because I feel they go together) are empathy and humility".

John listed five qualities. As stated it isn't easy to have just one. John said, "There are five qualities that are essential including honesty, competence, intelligence, forward looking and inspiring...but the most important of these is honesty. Without honesty, the other four qualities don't mean much!"

Ian said, "One quality seems unfair. A leader has multiple qualities I look for. A leader should be confident and knowledgeable but listen to ideas to build a powerful and impactful team".

"Integrity", was Rich's first response. He quickly followed up with another email with the following, "Honorable mention would be unselfishness".

Blake did respond with one quality. He said, "Attitude. The right attitude is not teachable, and it's not replaceable. Attitude affects everything a leader touches, or says or does. It doesn't matter what talents you have, what connections you have, what education you have. If you don't have the right attitude, you're going to fail."

Carol responded with "Integrity! Someone who is honest and open – can be trusted. Doesn't think of themselves but makes decisions that benefit the greater good!"

As you can see by these answers and responses, they have all taught and influenced me and I value their thoughts and insight. They are incredible people, and if in saying they that I highly respect them as leaders, as they are reading this, they are thinking they still have more to learn and want to strive to do more. They should take a moment and enjoy the fact that they are awesome user group leaders.

Again it was difficult to come up with one quality. Even the group of six had all pretty much come up with different answers except integrity. Honesty and integrity are extremely important and I am sure every one of these leaders has it in the top five if allowed to give more. It is important to be able to trust a leader. How can one follow if the leader is not looking out for interest of the group and the leader's intentions are trustworthy? We want to be able to respect and trust our leaders, even if you are a technical leader. Knowing that there are honest answers and even if it is "I don't know, let's find out" answer. Integrity and honesty are required leadership qualities. From these other qualities can be built.

In looking at the other qualities for a leader listed; humility, empathy, competence, intelligence, forward looking, inspiring, confidence, knowledgeable and attitude. Adding to the list is critical thinking, serving and enabling. It almost sounds like an impossible task to have all of these qualities, but don't forgot that is still a process and on-going learning. There are going to be qualities that come easy and others require the work and effort. As one looks at different leaders, there are also bad qualities that are recognized. Self-awareness will be able to determine if some of these bad qualities need to be suppressed. These qualities appear in leaders that we do not aspire to become. Self-awareness allows for the understanding what needs to be worked on and help examine if you are demonstrating these qualities.

Demonstrating Leadership

Practicing the qualities of leader is demonstrating leadership. It sounds easy right? For example, attitude is something you choose to show. A positive, can-do attitude goes a long way. Someone will be willing to work just as hard with a leader that is optimistic with a realistic plan, and even with a little work attitude will take it further. This does come in handy when dealing with the not always the fun stuff. There are budgets and cuts that might have to be made and decided on. An attitude demonstration how important these tasks are will make that process feel just as fun and needed as going out and meeting with the user community.

Another example would be demonstrating critical thinking. This is not only in the user group leadership role but also in the other roles with technology or business. There are plenty of problems to be solved as a user group; how to reach new members, dealing with generation gaps, conference attendance, new ways to present content, how to use social media, and many more.

Demonstrating leadership means doing the things that the user group leader does. Showing up to meetings and participating in discussions with the qualities of a leader. Confidence comes through when having the discussions and not being timid in presenting ideas. Everyone in the boardroom and volunteers has ideas to present, and if not presented with confidence it might not be accepted from the group.

A user group that has the leadership team working on the qualities and involved will have others being excited about volunteering and being part of the team. This provides another opportunity to inspire others to pick up the ball and run with it.

Serving the community is demonstrating leadership. A serving leader will look to serve the group in helping others succeed. This is what a user group does in helping members learn and be successful in their careers. It isn't just constant pouring yourself into the community and giving up on all of your needs, but it is realizing that there is value in serving the greater good. A serving leader is normally one that is recognized by the people being excited to be part of the community and being enabled to learn and network. It is not just letting everyone speak or not setting a standard for volunteers, but it is setting the expectations and demanding excellent quality of education. It sets a high bar for the volunteers, but also one that is obtainable. It gives them in turn an opportunity to serve the community back.

A serving leader is right there in the middle of it all. They just don't say do, but give expectations, leave room to discuss how something can be accomplished. Once it is decided what to do they are right in there helping out and supporting the effort and work. Just like with the different committees for the user group, this provides excellent ways to roll up your sleeves and participate with the group and serve the community.

Developing a succession plan for the board and leadership demonstrates how the important a transition is for continuity of the culture and governance. This puts the importance of the group ahead of the needs of leader. It too can plan out different ways for the user group volunteers to reach the board of directors if they choose to do so.

A leader demonstrates how to handle change and leads the process. Change is a constant and it is not always something easy to do, especially in a group that might be rooted in something the vendor did at one point or have quite a long history. Managing change from communication to doing by showing the way with the changes. Sometimes change is needed in the user group because of new technology or different needs of the community. There is always resistance to change because of history and how something has always been done this way. Social media has changed how we communicate, and something that might have been sent out or emailed might now just be a link or some of the face-to-face events might be a virtual offering now. Depending on these needs it might be difficult for some of the community to accept these changes, and the leaders have to demonstrate how that has helped them to communicate better or take advantage of more education. If the leaders and volunteers are helping out and showing the way the community is able to buy into more.

Leader Volunteer to President

This is a personal example and some of it was not easy nor did it go the way that I had expected. It is really just to give a brief story of the user group and how I went from being a member to a leadership role. It is something I didn't think was possible, and I am still very thankful for the opportunity. I have also pointed out that I am still supporting and involved in the user group and leading from a volunteer position because I enjoy being part of the community. Again, I am trying not to present this in a way of boasting, but just hopefully something that can be learned from. Even from recruiting volunteers, which will be discussed more in the next chapter, as you can see why I did some of the things I did to get more involved.

As a member I started to use Independent Oracle User Group (IOUG) as a resource for starting off as an Oracle DBA. I had work with other database platforms and was able to tap into the user community for the information related to Oracle and how to work on performance and other design considerations. In working on different projects, I continued to look for information and realized that there were ways to get more information and details from the vendor and provide feedback.

There was a volunteer leader that recognized the involvement I was seeking with the vendor and offered me a chance to develop on of the special interest groups. I gladly participated because it was the details and solutions I was looking for work, and I was able to work with a couple of other volunteers to build a community around the special interest group. Things I did as a leader in the special interest group was plan, set strategy recruit other volunteers and communicate with the community. We developed a plan to provide webinars and articles around these topics. We worked with the vendor to make sure we understood the roadmap of the product and how we could provide feedback. The schedule for the learning opportunities was communicated to the membership along with recruiting specifically for those members with this interest.

The work I had completed with the special interest group provided an overall charter for other groups to be developed. The structure, that was used to create the group helped, created governance and planning of additional special interest groups. It was then one of the board members plugging this structure into the overall strategy that board had to utilize the special interest groups. It was not just me but there great leaders on the board of directors being aware of the different options and enabling volunteers to be part of the development of the user group.

It was such a privilege to be asked to apply for the board. This again is a pattern of building a succession plan and finding good volunteers. There were different responsible areas for the board of directors. We each had our own portfolio, which allowed us to build committees from that and organized the governance to make sure that the different portfolios were connected to the overall strategy of the board. We did not always agree on the portfolios the different members had. There were some discussions within the boardroom. Debates and some other thoughts took place of which groups should be assigned. Outside the boardroom the decision made was supported.

In the years that I was on the board of directors we looked at the strategy and went back and forth to make sure we had the right strategies and goals set, which did change over the years because of changes in Oracle stack and changes in the community. These are good examples of what a board needs to do and the leadership of the board needs to be able to have the forward thinking and big picture for the community to be able to do this.

There were opportunities to learn from the other board members and their experiences. There were also formalized classes on leadership that were available. I took advantage of everything available to sharpen my skills and develop the important qualities of a leader. One of the most important things I learned from these classes was self-awareness because that taught me to look at my values and evaluate my motives. I could then realize how I was doing as a leader, and measure against people I was inspiring and allowing to also come along as volunteers and leaders in the community.

I had the privilege to attend the baseball game where the Cubs defeated the Cardinals in the 2015 Division Series. It had been awhile since I was able to attend a game or even sit and watch a game on TV. I did not just enjoy it because the Cubs won, even though that was pretty awesome, but I enjoyed watching the team how much they were involved in the game and how the fans fed off of that energy and participate with great enthusiasm as a key part of game. Previous Cub teams that I had seen play did not have the energy and did not have the excitement as being part of the game. This was an important game and there was already anticipation to win the division series, but the team had to work together to get here. The teamwork and eagerness of the team was outstanding. The Cub manager encouraged the fun and participation with the fans. He taught them what they needed to know. He let them know when things needed to change as the pitcher changed out and didn't wait too long to implement the change. They celebrated accomplishments. The leader of the team enabled his team to be excited to do what they liked to do and that was to play baseball.The last board meeting I was involved with, I was so excited how everyone was having great discussions, going back to strategies and evaluating the resources and what should be done. They were really a group excited about the IOUG and being part of the community. They had grown as leaders themselves and took succession planning seriously to make sure that it continued.

I hope that this brief description of the areas I was involved with in the user group helped of what might be important for leadership and areas that need to be developed by the board of directors. I am still speaking and recruiting volunteers for different areas to have people who are passionate about the community to remain involved.

Summary

A user group leader has a wide variety of things that they need to do to lead, support and inspire the community. Even as a leader, you do not have to go alone. There are others on the board of directors or other key volunteers that can help out to verify ideas support the tasks at hand and provide feedback about what the user group is doing and areas to work on.

There are many qualities of a leader with a few of them being discussed in this chapter. Integrity is a very important quality that a leader can demonstrate. A leader with integrity is one that is respected and will have people willing to follow. Leaders who are open and honest will communicate to the community in this way and gain the respect and followers.

There are some qualities that come with values, but others that can be developed and worked on just like skills that are needed. Self-awareness will help determine the qualities that you are already demonstrating and evaluate good versus bad qualities. The bad qualities are normally easy to determine because these are obvious in the people in charge that are not considered leaders.

Another quality is attitude and one that we might overlook, but having the right attitude being involved and excited about the user group is going to inspire others. A positive attitude is one that is contagious and the enthusiasm will spread throughout the group. Bringing energy and excitement to the community will draw others to volunteer and become more involved.

Learning and developing skills and qualities to become a leader requires practice, too, and demonstrating these skills in the user group and community. It does not have to be in a board role first but as part of a committee or another area that the user group can use the assistance. There are plenty of ways to demonstrate these qualities and they will be noticed either in the workplace or user group. These qualities are important for career building as we well as the involvement in the community.

This is definitely not an exhaustive list on how to demonstrate and all of the qualities that are needed. Remember leaders do have multiple qualities and it is not just one quality to work on. It is not an easy task, but well worth it when seeing the results and others enabled to learn and develop these skills too.

It is difficult to compare some of the same work we do to baseball, and the excitement of this game. To experience that on a regular basis within the user group would be amazing, and the leaders of the user group can enable the exhilaration and energy. Keeping up that level could get exhausting but it would not be boring. The leaders can inspire this type of team and community to be working toward a goal. It would be an awesome feeling to be part of a community that takes pride in what they do and provide. To be able to continue to do what we are good at and have a group of leaders that are developing talent in the membership and grow the community into champions.

■ ■ ■

CHAPTER 10

Qualities of a User Group Volunteer

Volunteers are critical to any community. There are several qualities that are important in a good volunteer. A volunteer is also developing leadership.

If this is new user group, the volunteers are out there, but they are just more difficult to find. They might not yet know of the user group or might have been working on something similar and it would have to be merged or partnered with in some way. But a new group also has more vendor support, so that is a good resource for recruiting volunteers.

Volunteers should be groomed for other roles. There are users who will want to volunteer for one simple task and that is enough for them to participate in the community in the way they want. Others will be interested in growing into more leadership roles and look to eventually be part of the board of directors. Part of succession planning should include qualities that you look for in volunteers. It does not have to describe a skill set because a needed skill set might change as the user group develops or needs to change.

This chapter will look at some of the tasks for volunteers. There are development paths and ways to communicate to key volunteers about potential that they have in the user group.

What Does a User Group Volunteer Do?

Some people would prefer to stay far away from speaking and can volunteer for many other areas. There are several different activities that would fit under these categories:

- **Support**—Understand that the user group is for them and support the activities that are provided.

- **Provide feedback**—Volunteers have a different perspective and can feedback about the user group activities.

- **Support events**—There are speakers, planners, coordinators, and positions to follow up afterward, all of which could be handled by volunteers.

- **Committee involvement**—Committees are the main ways for planning for the user group. Therefore, volunteers need to be there to help come up with the ideas and planning. They can then help with the different tasks that arise.

- **Recruit**—Using volunteers to recruit new volunteers is a great idea. Current volunteers can talk about time commitments, activities, and the excitement that they have being part of the user group.

- **Vendor liaison**—The vendor will still like feedback from the customers and the users are their customers. Pulling a few key members and volunteers in a room to talk to issues or enhancements is beneficial for the user and vendor.

These are just some of the things that volunteers do. It is important to have a list of activities and tasks available when reaching out to volunteers so that they can see how they can participate right away. Engagement of the volunteers is should be very quick because as people start to be interested, the tasks and volunteer jobs will draw them in to the activities. The connections are going to help bring things together when there is energy around volunteering.

Member to Volunteer

A user group can be a fairly large community. There are going to be those members that are more engaged than others and others who will just look in for information and resources.

It is difficult to pull volunteers out of the member community without some networking time. Either there is a current volunteer that knows the member and encourages involvement or an event makes it easier to bring a user on board. It also helps so they know what they are part of and start to see the community and how valuable it is to the group.

It is helpful to have a website with volunteer opportunities, and registrations need to be followed up on; this would be a great task for a volunteer. Figure 10-1 shows the progression from member to volunteer.

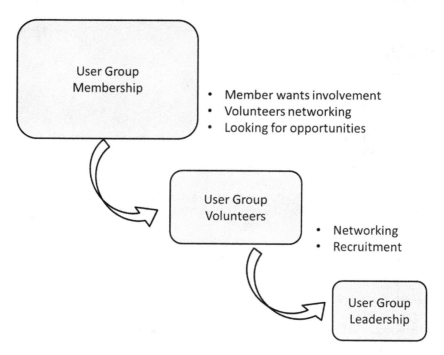

Figure 10-1. *Progression from member to volunteer*

There is interest either from the member or another volunteer to encourage and give examples of how to become more involved. I have seen this so many times, just a little bit of time and real (remember integrity and honesty here) interest from a speaker, a leader or a volunteer goes a long way for member to become more interested in the user group and want to volunteer. They are looking for opportunities so the volunteers should be ready to give them ideas. This applies to the volunteers that want more of a leadership role, and the board will be working with volunteers to encourage and engage them to become more involved. The board and leadership can also be reaching out to the members. It is definitely not a strict hierarchy because we are all users coming from the same place and are all working to develop skills and work together. Board members might have even more of an influence on the members becoming volunteers. Part of the leadership is sharing the story, explaining the benefits, and talking about ways to be involved. There are definite benefits to be part of the user group and spend time networking and learning.

Qualities Outlined

The qualities to look for in a volunteer might include:

- Integrity
- Confidence
- Competence

- Attitude

- Passion

- Professionalism

- Team Player

- Energy

Integrity is needed to be able to have the same honest and open communication with the membership and other volunteers. Having experts in the community to speak and work with others is why confidence and competence are also needed for volunteers as well as the leaders. Some of the confidence and even competence in some of the areas might still need to be developed, so that is where professionalism comes in. The respect and working through presenting or even issues from a committee demands professionalism.

The right attitude for a volunteer is just as important for a leader. The attitude will make it easier to be energized and ready to participate in the user group and the volunteer activities. People with passion are going to work with the group and are excited to be a part of the community. They want to learn and develop, and they want to be involved. The right attitude and passion will allow for many other qualities to be developed and build them into leadership roles.

Energy is required. This is a volunteer position, which is something that is done outside of the normal working hours, and with technology, we are not even sure what normally working hours are any more. This does not mean saying yes to everything, but it does mean being able to have the energy to be involved in the committee and being able to participate. Ideas, input, and work needs to be done, so one does need to have energy to work through the meetings and other tasks.

The qualities are not far off of those of a leader. Integrity is very important. It is a core value that attracts others to want to work with that person and trust enough to be inspired.

Is a Volunteer a Leader?

Yes, a volunteer is a leader and anyone can be a leader. They don't have to have the title. They have to be able to enable and inspire others. They serve others and are constantly learning. This definitely sounds like a volunteer to me, wait sounds like a leader too.

Volunteers are going to be leaders in the community. They have the responsibility to the user group to perform the activities and tasks they commit to. They are recruiting others and even in the different positions they are in, they should be planning for succession as well.

This is something that I would do at events, especially when there is someone with a good question or interested in a topic at conference. I mention, that would be a great topic for you to speak on next time. I didn't leave it at that, because a causal comment like that can be dismissed. I would follow up and see if they needed any help or had thought more about what they were going to speak on. I was excited to attend the next conference because there were some new sessions and speakers I was very interested in hearing.

A volunteer is a leader in their areas of expertise and a leader in the community. They will want to continue to develop leadership skills and use the user group experience to practice these skills.

Summary

The qualities of a user group volunteer are very similar to those of a leader. Some of these areas are still being developed and the user group should provide them ways to practice and participate in activities that build these skills.

There are suggestions for activities for volunteers and there are several tasks that can engage a volunteer. Part of planning and creating a strategy would have a list of these tasks to be able to involve those interested. This helps for members to know how they can volunteer and increase their activity in the community.

The user community is ideally for the volunteers to be able to network and learn from each other. Conferences and events would not be possible without the volunteers planning, reading papers and selecting topics and speakers. Articles, webinars, and blogs are ways that other volunteers can provide information throughout the year.

With so many books on leadership, there are plenty of opportunities to continue to learn and develop leadership skills. The user group community is a perfect place to not only expand their knowledge in the vendor products but to also practice and develop leadership skills. The members in a user group community have opportunities to grow into new roles and leaders. The user group has so many benefits for the members and vendor, and starting a new user group or becoming a leader in an existing one is an exciting adventure. I would not have had the opportunity to meet community members around the world, speak in cities that I didn't think of visiting, and continue to have learned so much. I am passionate about what I do, and having the privilege of being able to be a leader in a user group increases that passion and enthusiasm. I am a lifelong learner and have still many skills to continue to develop.

Leadership doesn't come from the title or role, but the qualities being developed and how you inspire and enable others to participate and be successful in their careers. Being part of a user group community is an excellent tool for career advancement and leadership skills can be used in the community and many different aspects a job. I hope that this book encourages those developing the user group community to continue and as a volunteer you receive quite a bit more than what you put into it.

Index

V

W, X, Y, Z

Get the eBook for only $5!

Why limit yourself?

Now you can take the weightless companion with you wherever you go and access your content on your PC, phone, tablet, or reader.

Since you've purchased this print book, we're happy to offer you the eBook in all 3 formats for just $5.

Convenient and fully searchable, the PDF version enables you to easily find and copy code—or perform examples by quickly toggling between instructions and applications. The MOBI format is ideal for your Kindle, while the ePUB can be utilized on a variety of mobile devices.

To learn more, go to www.apress.com/companion or contact support@apress.com.

GPSR Compliance
The European Union's (EU) General Product Safety Regulation (GPSR) is a set
of rules that requires consumer products to be safe and our obligations to
ensure this.

If you have any concerns about our products, you can contact us on

ProductSafety@springernature.com

In case Publisher is established outside the EU, the EU authorized
representative is:

Springer Nature Customer Service Center GmbH
Europaplatz 3
69115 Heidelberg, Germany